Series in Biomedical Engineering

Series editor

Joachim H. Nagel, Stuttgart, Germany

This series describes the applications of physical science, engineering and mathematics in medicine and biology. The books are written for graduate students and researchers in many disciplines including medical physics, biomedical engineering, radiology, radiotherapy and clinical research. The Series in Biomedical Engineering is the official book series of the International Federation for Medical and Biological Engineering.

The IFMBE is an association of constituent societies and organizations which was established in 1959 to encourage and promote international collaboration in research and practice of the profession as well as in the management of technology and the use of science and engineering in medicine and biology for improving health and quality of life. Its activities include participation in the formulation of public policy and the dissemination of information through publications and forums. The IFMBE as the only international organization and WHO/UN accredited NGO covering the full range of biomedical/clinical engineering, healthcare, and healthcare technology management, represents through its 50 national and international member societies more than 120.000 professionals involved in the issues of improved health care delivery.

The IFMBE is associated with the International Union for Physical and Engineering Sciences in Medicine (IUPESM). Through the IUPESM, the IFMBE is a member of the International Council for Science (ICSU). http://www.ifmbe.org

More information about this series at http://www.springer.com/series/7752

Almir Badnjević · Mario Cifrek
Ratko Magjarević · Zijad Džemić
Editors

Inspection of Medical Devices

For Regulatory Purposes

 Springer

Editors
Almir Badnjević
Medical Devices Verification Laboratory
Verlab Ltd Sarajevo
Sarajevo
Bosnia and Herzegovina

Mario Cifrek
Faculty of Electrical Engineering
   and Computing
University of Zagreb
Zagreb
Croatia

Ratko Magjarević
Faculty of Electrical Engineering
   and Computing
University of Zagreb
Zagreb
Croatia

Zijad Džemić
Institute of Metrology of Bosnia
   and Herzegovina
Sarajevo
Bosnia and Herzegovina

ISSN 1864-5763
Series in Biomedical Engineering
ISBN 978-981-10-6649-8 (hardcover)   ISBN 978-981-10-6650-4 (eBook)
ISBN 978-981-13-4923-2 (softcover)
https://doi.org/10.1007/978-981-10-6650-4

Library of Congress Control Number: 2017954478

Printed on acid-free paper

This Springer imprint is published by Springer Nature
The registered company is Springer Nature Singapore Pte Ltd.
The registered company address is: 152 Beach Road, #21-01/04 Gateway East, Singapore 189721, Singapore

Contents

Introduction

Almir Badnjević, Mario Cifrek, Ratko Magjarević and Zijad Džemić

Abstract Inspection of Medical Devices—for regulatory purposes is an overview of the expanding and exciting field of Medical Devices in which the reader will find a modern presentation of the relevant aspects of inspection of medical devices as part of the legal metrology system.

Around the world, there are a variety of health care systems, each with its own characteristics and organizational structure according to national resources, requirements and needs. Bearing this in mind it is very difficult to give a general definition for the health care system. Basically, it can be defined as a normatively accepted system of society and government in protecting and improving the population health, with all system factors affecting organized and constantly evolving as part of a general country social system. Each health care system consists of medical institutions which in addition to personnel and infrastructure must possess the necessary equipment in order to perform the correct diagnosis and treatment of their patients. In addition to the knowledge and experience of medical doctors, in patient diagnosis and treatment, it is necessary to have the correct and tested medical apparatus.

Diversity and innovativeness of medical devices, as a result of evolving field of biomedical engineering, significantly contribute to improvement in quality and efficiency of healthcare services. European commission define the term "medical device" as any instrument, apparatus, appliance, software, material or other article, whether used alone or in combination, including the software intended by its

A. Badnjević (✉)
Medical Devices Verification Laboratory, Verlab Ltd Sarajevo,
Sarajevo, Bosnia and Herzegovina
e-mail: almir.badnjevic@verlab.ba

M. Cifrek · R. Magjarević
Faculty of Electrical Engineering and Computing Zagreb,
University of Zagreb, Zagreb, Croatia

Z. Džemić
Institute of Metrology of Bosnia and Herzegovina, Sarajevo, Bosnia and Herzegovina

© Springer Nature Singapore Pte Ltd. 2018
A. Badnjević et al. (eds.), *Inspection of Medical Devices*, Series in Biomedical
Engineering, https://doi.org/10.1007/978-981-10-6650-4_1

1

manufacturer to be used specifically for diagnostic and/or therapeutic purposes. Covering a wide range of products, from bandages to the most sophisticated life-supporting products used in diagnosis, prevention, monitoring, and treatment of diseases proper functionality of medical devices is crucial. In particular, it is important in life critical situations, when doctors have no more than 10 min to make a decision according to diagnosis based on readings of medical devices. Unfortunately, between 40,000 and 80,000 patients around the world, die due to the malfunctioning of medical apparatus and over 10,000 patients get seriously injured. Due to these facts, withdrawals of series of medical devices from the market by its manufacturers have been registered in the past.

The aspect of safety of medical apparatus in health care systems around the world is regulated by different agencies or by applying international managing standards for health care institutions, which ensure that safety of medical devices is checked once a year. On the other hand, the aspect of safeness of medical apparatus in a health care system for most countries is left to manufacturers or distributors of medical equipment, allowing them a certain kind of monopoly.

The majority of Medical devices are under authorized preventive and corrective maintenance in different ways. In some health institutions preventive service is scheduled annually, while in some institutions four times per year, with a great impact for the budget of the health care institution and for the overall health care system. This is another kind of previously mentioned monopoly.

As a part of preventive service, an authorized service center performs also certification of apparatuses. A certification process report is usually a work order document. This document only reports the result of the certification: whether the device passed or failed. The work order document contains neither any information about device output values measurement, nor the reference to the certification standard.

In order to comply with the recommendations of the World Health Organization (WHO) and international standards for medical devices, it is necessary to apply the law in the field of metrology for medical apparatus. In that way every medical device is considered as a part of legal metrology which is regularly checked for deviations of output values. These deviations must not exceed the defined limits in order for the medical device to be safe for use on patients. International standards and norms concerning medical devices are Medical Device Directive 93/42/EEC, Medical Electrical Equipment ISO 60601 and Safety Testing of Medical Devices ISO 62353. In addition, the international standard ISO 17020 defines the system of competence management of laboratories that deals with the inspection.

In addition to knowledge and experience of medical doctors, correct diagnosis and appropriate patient treatment largely depend on accuracy and functionality of medical devices. In a large number of serious medical situations proper functionality of medical devices is crucial for patients. Therefore, it is necessary to carry out as strict and independent testing of functionalities of medical devices as possible and to obtain the most accurate and reliable diagnosis and patient treatment.

This book presents the experience in directives and standards regarding Medical Devices that highlight the necessity of introducing metrology into medicine and defining standard regulations for inspections of medical devices. As it has been previously done for other kinds of devices that are under jurisdiction of the National Metrology Institutes, this book provides a foundation for the introduction of medical devices into the legal metrology system with precisely defined units of measurement, their ranges and errors.

Medical devices which are covered in this book and suggested as part of the legal metrology system include ECG devices, defibrillators, patient monitors, respirators, anesthesia machines, dialysis machines, pediatric and neonatal incubators, therapeutic devices, and infusion as well as perfusion pumps. Furthermore, standard inspection regulations for the aforementioned medical devices are also defined. Additionally, there is explanation for what is necessary to establish an independent laboratory for the inspection of medical devices under the ISO 17020 standard.

With the introduction of medical devices into the legal metrology system and with the establishment of a fully operational medical devices inspection laboratory, it is expected that the reliability of medical devices in diagnosis and patient care will increase and that the costs of the health care system in countries will be reduced.

This book was compiled to be an indispensable resource for professionals working directly or indirectly with medical devices, national metrology institutes, institutes of accreditation and institute of standardization. Just as importantly, it was organized for graduate and postgraduate students in biomedical engineering, electrical engineering, mechanical engineering, hospital engineering and medical physics.

The editors would like to thank the authors for their fruitful, successful and collegial cooperation. It was a pleasure for us to collect views from the different fields of medical devices and bring them together in the book.

Regulations and Directives—Past, Present, Future

Haris Memić, Almir Badnjević and Zijad Džemić

Abstract Regulations are binding acts which are obligatory in the European Union. All members of the European Union must apply Regulations. On the other hand, there are Directives, legislative acts which represent the base setting the goal which has to be achieved by EU countries for a specific area. Each EU country has individual national laws in order to achieve that goal. Directives which regulate harmonized products in the EU are known as New Approach Directives. One of the Directives which belong to the group of the New Approach is Directive 93/42/EEC on medical devices. The integral part of a Directive is Harmonised Standard which serves Manufacturers, other economic operators, or conformity assessment bodies to demonstrate that products, services, or processes comply with relevant EU legislation. Conformity assessment is a process that is performed by the manufacturer in order to demonstrate if all specific requirements related to the product have been met. Conformity assessment is provided by a competent body (notified bodies) and differs for different classification of medical devices. There are different approaches in conformity assessment of medical devices in the EU and USA which are described in this chapter. Many European countries have recognised the importance of metrology and its influence in providing services which will ensure accurate and precise measurements of medical devices that have a measurement function.

1 Introduction

EU legislation is divided into two levels with primary legislation embodied by the treaties, and secondary legislation given in the form of regulations, directives and decisions which are used to implement the policies set out within the treaties [1].

H. Memić (✉) · Z. Džemić
Institute of Metrology of Bosnia and Herzegovina, Sarajevo, Bosnia and Herzegovina
e-mail: haris.memic@met.gov.ba

A. Badnjević
Medical Devices Verification Laboratory, Verlab Ltd Sarajevo, Sarajevo,
Bosnia and Herzegovina

© Springer Nature Singapore Pte Ltd. 2018
A. Badnjević et al. (eds.), *Inspection of Medical Devices*, Series in Biomedical
Engineering, https://doi.org/10.1007/978-981-10-6650-4_2

Secondary legislation is made by the EU institutions. It is the third major source of Community law after the treaties (primary legislation) and international agreements. It comprises:

- binding legal instruments (regulations, directives and decisions)
- non-binding instruments (resolutions, opinions).

EUR-Lex provides free access to EU law and other documents considered to be public. The content on the official website is available in 24 official languages of the European Union.

This chapter has the purpose to describe the legislation used in EU in the field of medical devices, and the approaches to be followed by producers (manufacturer) of medical devices in order to be approved and placed in the EU market.

In order to increase safety in the production of medical devices, manufacturers have to follow the relevant legislation. This legislation in the EU is given through the directives and regulations followed by appropriate harmonized standards set out in the directives. In addition to the stated documentation, there are also other acts of European Union Law.

The description and meaning of legal acts in accordance with EU law is given bellow [2]:

A "**regulation**" is a binding legislative act. It must be applied in its entirety across the EU.

A "**directive**" is a legislative act that sets out a goal that all EU countries must achieve. However, it is up to the individual countries to devise their own laws on how to reach these goals.

A "**decision**" is binding on those to whom it is addressed (e.g. an EU country or an individual company) and is directly applicable.

A "**recommendation**" is not binding. When the Commission issued a recommendation that EU countries' law authorities improve their use of videoconferencing to help judicial services work better across borders, this did not have any legal consequences. A recommendation allows the institutions to make their views known and to suggest a line of action without imposing any legal obligation on those to whom it is addressed.

An "**opinion**" is an instrument that allows the institutions to make a statement in a non-binding fashion, in other words without imposing any legal obligation on those to whom it is addressed. An opinion is not binding. It can be issued by the main EU institutions (Commission, Council, Parliament), the Committee of the Regions and the European Economic and Social Committee. While laws are being made, the committees give opinions from their specific regional or economic and social viewpoint.

Review of EU legislation [3] can be made via the official website of EU which is especially dedicated to this issue, http://eur-lex.europa.eu.

An integral part of the directives are harmonised standards.

A **harmonised standard** [4] is a European standard developed by one of the recognised European Standards Organisation: CEN [5], CENELEC [6], or ETSI [7]. It is created following a request from the European Commission to one of these organisations. Manufacturers, other economic operators, or conformity assessment bodies can use harmonised standards to demonstrate that products, services, or processes comply with relevant EU legislation.

The references of harmonised standards must be published in the Official Journal of the European Union. The purpose of this website is to provide access to the latest lists of references of harmonised standards and other European standards published in the Official Journal of the European Union (OJEU).

Medical devices

This chapter describes the approach of placing medical devices in the EU market, the process of approval in accordance with appropriate EU directive, the role of notified bodies, processes of conformity assessment, comparison with approach used in USA and the role of medical devices under the frame of Legal Metrology.

A "**medical device** [8]" is any instrument, apparatus, appliance, material or other item, whether used separately or combined with another device, including the software necessary for its proper application intended by the manufacturer intended for human use for the purpose of:

- diagnosis, prevention, monitoring, treatment or alleviation of disease,
- diagnosis, monitoring, treatment, alleviation of/or compensation for an injury or handicap,
- investigation, replacement or modification of the anatomy or of a physiological process,
- control of conception, and which does not achieve its principal intended action in or on the human body by pharmacological, immunological or metabolic means, but which may be assisted in its function by such means.

2 Placing of Medical Devices in the Market of EU in Accordance with EU Legislation

Before the placing of a medical device in the market, implying that it is ready for use, the device has to be approved by competed body providing conformity assessment of a tested subject with appropriate reference. There are different recognized approaches in the process of approval.

In most countries medical devices are categorized based on the risks associated with their use, and the approval process varies by category [9]. In some countries like the United Kingdom, manufacturers of low-risk devices may register with the government agency and simply declare that the devices meet the requirements to be approved, but devices classed as higher risk must undergo a more detailed review performed by a notified body. In most cases within the EU the approach of putting the medical device in the market is similar, but it differs if compared with the approach used in the USA, which will be described later in this paper.

Medical Device Approval Process in EU

Medical devices must comply with the rules established by EU directives related to medical devices prior to being put in the market and/or put into service in the EU, the European Economic Area, or Switzerland. At EU level, there is no centralized approach similar to that in the United States.

The European Medicines Agency [10], unlike the Federal Drug Administration in the United States, is not involved in the approval process of medical devices. Manufacturers, prior to placing their devices in the market, are required to determine the classification of a device, based on the risk factors associated with each device, and then to apply the appropriate conformity route. Medical devices are assessed for efficiency and safety by notified bodies, generally private organizations, staffed by experts and certified by the EU Member States. In the final stage, medical devices, with some exceptions for such things as custom made devices and devices intended for clinical investigation, are given a CE marking, which ensures that medical devices are in conformity with EU rules and are ready to be marketed.

3   EU Legislation in the Field of Medical Devices

Legislation in the field of medicine in the European Union, i.e. legislation relating to medical devices, is based on Council Directive **93/42/EEC** of 14 June 1993. This directive further relies on the following directives [10]:

- Directive **93/68/EEC** (CE Marking);
- Directive **98/79/EC** of the European Parliament and of the Council of 27 October 1998 on in vitro diagnostic medical devices;
- Directive **2000/70/EC** of the European Parliament and of the Council of 16 November 2000 amending Council Directive 93/42/EEC as regards to medical devices incorporating stable derivates of human blood or human plasma;
- Directive **2001/104/EC** of the European Parliament and of the Council of 7 December 2001 amending Council Directive 93/42/EEC concerning medical devices;
- Directive **2007/47/EC** of the European Parliament and of the Council of 5 September 2007 amending Council Directive 90/385/EEC on the approximation of the laws of the Member States relating to active implantable medical devices, Council Directive 93/42/EEC concerning medical devices and Directive 98/8/EC concerning the placing of biocidal products on the market.

In the past, one more directive was in force, but it has been repealed as of January 1, 1995. This was the Council Directive **76/764/EEC** on the approximation of the laws of the member states on clinical mercury-in-glass, maximum reading thermometers.

As mentioned in the introduction, beside directives, there are a great number of written standards used in manufacturing of medical devices.

In the field of medical devices, there are ca. 200 standards which have been issued under the European Committee for Standardization CEN (without revision), and close to 100 standards issued under the European Committee for Electrotechnical Standardization CENELEC (without revision) [11]. Taking into account such a large number of standards regulating the requirements for a particular product group, it is easy to conclude that this area of manufacturing presents an area where most attention is paid in relation to the safety of the product itself and its users.

EU Directive 93/42/EEC

The most comprehensive application in the field of medical devices of all directives mentioned in the text above has been the Directive 93/42/EEC which is the basis for the production of medical devices, and other directives complementing this field. This Directive was first published in the year 1993 and till 2007 it has been reviewed several times. Revisions of Directive 93/42/EEC were published in the following years:

- 12/07/1993
- 07/12/1998
- 13/12/2000
- 10/01/2002
- 20/11/2003
- 11/10/2007.

Directive 93/42/EEC belongs to the group of the New Approach directives, meaning that the products regulated by this Directive fall within the group of harmonized products in the EU. The application of the directive is a legal requirement for all manufacturers of medical devices in EU.

New Approach is a regulatory technique used for removing technical barriers to trade in Europe. For this purpose a common set of harmonised technical regulations were implemented in the member countries of the EU and in some non-member countries as well. The number of directives based fully or partially on the principles of New Approach is 30 [12].

Each New Approach directive stipulates the essential requirements, such as product safety or reliability which are to be satisfied by the product. Products, which satisfy these conditions, may be sold freely inside and among the European countries. New Approach directives have a binding effect as to the goals set up in them, but the member states can choose the methods to reach these goals in their national legislation.

In accordance with the directives of the New Approach, the manufacturer represents a natural or legal person responsible for the design or manufacture and release of the product in the EU market under his/her name or his/her trade name.

The New Approach directives apply to products which are intended to be released on the EU market. In order to put the product on the market, it has to be assessed for conformity with the requirements defined by the specific directive. If the product, which is a subject of the conformity assessment, must satisfy the requirements of other directives, then the conformity assessment has to also be performed in accordance with those other directives.

In relation to the conformity assessment the manufacturer has certain responsibilities, reflected depending on the applied procedures. The manufacturer must take all necessary measures to ensure that the production process satisfy conformity of the product in order to set the CE mark on the product, which includes drawing technical documentation and creating EC[1] declaration of conformity. Depending on the directive, it is possible to claim from the manufacturer to submit products for testing and certification by a third party (usually a notified body) or to certify quality system by a notified body.

4 Notified Bodies for Conformity Assessment According to Directive 93/42/EEC

In the phase of the production, Medical devices are subject of compliance with certain directives and standards. Currently, in Europe, the most applied EU Directives for the production of measuring devices are 90/385/EEC, 93/42/EEC and 98/79/EEC (including the revision of 2007/47/EEC). The greatest responsibility in the production of medical devices holds the manufacturer himself, who has to ensure that the product meets the applicable legal requirements and, on the other hand, there is a notified body for conducting conformity assessment appointed by EU governments and with the obligation to validate and ensure that the product fulfil all the relevant requirements prescribed by the relevant directives.

At the moment, NANDO [13] database comprises 58 registered bodies competent to perform conformity assessment of medical devices in accordance with Directive 93/42/EEC. The highest number of notified bodies comes from Germany, 11 of them to be exact. Notified bodies perform tasks related to conformity assessment procedures, according to the applicable harmonized technical legislation when it requires the participation of a third party. Notified bodies may offer their services in the EU, but also to the third countries.

---

[1]As part of conformity assessment, the manufacturer or the authorised representative must draw up a Declaration of conformity (DoC). The declaration should contain all information to identify:

- the product
- the legislation according to which it is issued
- the manufacturer or the authorised representative
- the notified body if applicable
- a reference to harmonised standards or other normative documents, where appropriate.

5 Conformity Assessment of Medical Devices

Conformity assessment is a process that is performed by the manufacturer in order to demonstrate if all specific requirements related to the product have been met. The product itself is subject to conformity assessment in the stages of design and production. EU directives require the conduction of the process composed of one or two modules of conformity assessment. Conformity assessment of medical devices in accordance with Directive 93/42/EEC shall be carried out in accordance with the classification of the device itself.

In accordance with the directive 93/42/EEC, all medical devices are divided into four accuracy classes of increased risk. Accuracy classes of medical devices and related conformity assessment procedure in accordance with directive 93/42/EEC are given in Table 1 [14].

Products must be designed and manufactured in such a way that when used under proscribed conditions and for the intended purpose, they do not compromise neither patient safety, nor security and health of other users.

Medicinal products for the purpose of conformity assessment procedures, and for the given level of risk for the user, are divided into:

- Class I—medical devices with a low level of risk for the user,
- Class IIa—medical devices with a higher degree of risk for the user,
- Class IIb—medical devices with a high degree of risk for the user,
- Class III—medical devices with the highest degree of risk for the user.

This classification is done in accordance with the Annex IX of Directive 93/42/EEC. Classification rules are based on the sensitivity of the human body, taking into account the possible risks associated with the technical production and manufacture of medical devices.

Products must be in compliance with the essential requirements set out in Annex I of Directive 93/42/EEC which applies to them, taking into account the purpose of use of these products.

Table 1 Accuracy classes of medical devices and related conformity assessment procedure

Conformity assessment procedures	Classes					
Annexes	I	I sterile	I measure	IIa	IIb	III
II (+ section 4)						√
II (− section 4)		√	√	√	√	
III					√	√
IV		√	√	√	√	√
V		√	√	√	√	√
VI		√	√	√	√	
VII	√	√	√	√		

Table 2 Annexes to be applied depending on the class of medical device

Annex II	EC declaration of conformity (full quality assurance system) with or without examination of the design of the product
Annex III	EC type examination
Annex IV	EC verification
Annex V	EC declaration of conformity (production quality assurance)
Annex VI	EC declaration of conformity (product quality assurance)
Annex VII	EC declaration of conformity

Depending on the accuracy class of increased risks of medical devices, different procedures of conformity assessment in accordance with the Annexes of Directive can be applied. Annexes to be applied depending on the class are given in Table 2.

Technical documentation relating to the products of class IIa and IIb shall be examined by the notified body based on the program of a representative sample in the context of Annex II, V and VI of Directive 93/42/EEC.

As found in NANDO database, under Directive 93/42/EEC conformity assessment can be performed on the following products:

- General non-active, non-implantable medical devices
- Non-active devices for anaesthesia, emergency and intensive care
- Non-active devices for injection, infusion, transfusion and dialysis
- Non-active orthopaedic and rehabilitation devices
- Non-active medical devices with measuring function
- Non-active ophthalmologic devices
- Non-active instruments
- Contraceptive medical devices
- Non-active medical devices for disinfecting, cleaning, rinsing
- Non-active devices for in vitro fertilisation (IVF) and assisted reproductive technologies (ART)
- Non-active medical devices for ingestion
- Non-active implants
- Non-active cardiovascular implants
- Non-active orthopaedic implants
- Non-active functional implants
- Non-active soft tissue implants
- Devices for wound care
- Bandages and wound dressings
- Suture material clamps
- Other medical devices for wound care
- Non-active dental devices and accessories
- Non-active dental equipment and instruments
- Dental materials
- Dental implants
- General active medical devices

- Devices for extra-corporal circulation, infusion and haemopheresis
- Respiratory devices, devices including hyperbaric chambers for oxygen therapy, inhalation anaesthesia
- Devices for stimulation or inhibition
- Active surgical devices
- Active ophthalmologic devices
- Active dental devices
- Active devices for disinfection and sterilisation
- Active rehabilitation devices and active prostheses
- Active devices for patient positioning and transport
- Active devices for in vitro fertilisation (IVF) and assisted reproductive therapy (ART)
- Software
- Medical gas supply system and parts thereof
- Devices for imaging
- Imaging devices utilising ionizing radiation
- Imaging devices utilising non-ionizing radiation
- Monitoring devices
- Monitoring devices of non-vital physiological parameters
- Monitoring devices of vital physiological parameters
- Devices for radiation therapy and thermo therapy
- Devices utilising ionizing radiation
- Devices utilising non-ionizing radiation
- Devices for hyperthermia/ hypothermia
- Devices for (extracorporeal) shock-wave therapy (lithotripsy).

Horizontal Technical Competences

- Medical devices incorporating medicinal substances according to Directive 2001/83/EC
- Medical devices utilising tissues of animal origin, including Regulation 722/212 (Directive 2003/32/EC up to 28.08.2013)
- Medical devices incorporating derivates of human blood, according to Directive 2000/70/EC, amended by Directive 2001/104/EC
- Medical devices referencing the Directive 2006/42/EC on machinery
- Medical devices in sterile condition
- Medical devices utilising micromechanics
- Medical devices utilising nanomaterials
- Medical devices utilising biological active coatings and/or materials or being wholly or mainly absorbed
- Medical devices incorporating software/utilising software/controlled by software.

In the "Blue Guide" [15] on the implementation of EU rules on the products, the modules that are used to carry out conformity assessment are listed. In total, there are eight modules labelled with letters A through H. Modules specify responsibilities of the manufacturer and level of participation of in-house accredited bodies or notified bodies for conformity assessment. It is important to notice that an in-house body cannot act as a notified body, but they must demonstrate the same technical competence and impartiality to external bodies through accreditation as notify bodies. Conformity assessment bodies which carry out the assessment in accordance with modules that are shown in flow-chart (below) for the medical devices must meet the implementation of certain international standards (EU standards), or a combination of them.

The respective standards are: EN ISO/IEC 17020 (Requirements for the operation of various types of bodies performing inspection), EN ISO/IEC 17021 (Requirements for bodies providing audit and certification of management systems) and EN ISO/IEC 17065 (Requirements for bodies certifying products, processes and services) which are focused on criteria of conformity assessment, while standard EN ISO/IEC 17025 (General requirements for the competence of testing and calibration laboratories) deals with aspects of testing and calibration.

Bodies performing conformity assessment shall be accredited by the national accreditation body for specified standards. Accreditation implies confirmation of competences of the third party to an authority that may perform conformity assessment, respectively conformity with the requirements of applicable standards and additional requirements for the subject matter. Not all conformity assessment notified bodies from the NANDO base for the Directive 93/42/EEC have accredited their services. Some of them have confirmed their competences (they have been assessed) in accordance with Commission Implementing Regulation (EU) No 920/2013 [16]. In this case, assessment is performed by designating authorities [16] and joint assessment teams, which is built from the pool of national expert assessors made available from all of the designating authorities.

Setting out the requirements for accreditation and market surveillance relating to the marketing of products is done by Regulation 765/2008/EC [17], which should be seen as a complementary to Decision 768/2008/EC (on a common framework for the marketing of products). Accreditation provides authoritative statement about technical competence of bodies whose task is to ensure conformity with the applicable requirements.

The following Table 3 show modules in accordance with the "Blue guide", their description and related standards which have to be applied (or combination) in order to fulfil requirements of conformity assessment (Fig. 1).

Table 3 Overview of modules

Module	Description of the module	Applicable standards
A Internal production control	Covers both design and production The manufacturer himself ensures the conformity of the products to the legislative requirements (no EU-type examination)	EN ISO/IEC 17025 (+ ability to decide on conformity), or EN ISO/IEC 17020, EN ISO/IEC 17025 to be taken into account for testing required,
A1 Internal production control plus supervised product testing	Covers both design and production. A + tests on specific aspects of the product carried out by an in-house accredited body or under the responsibility of a notified body chosen by the manufacturer	or EN ISO/IEC 17065, EN ISO/IEC 17025 to be taken into account for testing required
A2 Internal production control plus supervised product checks at random intervals	Covers both design and production. A + product checks at random intervals carried out by a notified body or an in-house accredited body	
B EU-type examination	Covers design. It is always followed by other modules by which the conformity of the products to the approved EU-type is demonstrated. A notified body examines the technical design and/or the specimen of a type and verifies and attests that it meets the requirements of the legislative instrument that apply to it by issuing an EU-type examination certificate. There are 3 ways to carry out an EU-type examination: (1) production type, (2) combination of production type and design type and (3) design type	EN ISO/IEC 17020, EN ISO/IEC 17025 to be taken into account for testing required, or EN ISO/IEC 17065, EN 17025 to be taken into account for testing required
C Conformity to EU-type based on internal production control	Covers production and follows module B Manufacturer must internally control its production in order to ensure product conformity against the EU-type approved under module B	EN ISO/IEC 17025 (+ ability to decide on conformity), or EN ISO/IEC 17020, EN ISO/IEC 17025 to be taken into account for testing required, or

(continued)

Table 3 (continued)

Module	Description of the module	Applicable standards
C1 Conformity to EU-type based on internal production control plus supervised product testing	Covers production and follows module B. Manufacturer must internally control its production in order to ensure product conformity against the EU-type approved under module B. C + tests on specific aspects of the product carried out by in-house accredited body or under the responsibility of a notified body chosen by the manufacturer (*)	EN ISO/IEC 17065, EN ISO/IEC 17025 to be taken into account for testing required
C2 Conformity to EU-type based on internal production control plus supervised product checks at random intervals	Covers production and follows module B. Manufacturer must internally control its production in order to ensure product conformity against the EU-type approved under module B. C + product checks at random intervals tests on specific aspects of the product carried out by a notified body or in-house accredited body	
D Conformity to EU-type based on quality assurance of the production process	Covers production and follows module B. The manufacturer operates a production (manufacturing part and inspection of final product) quality assurance system in order to ensure conformity to EU type. The notified body assesses the quality system	EN ISO/IEC 17021 (+ product related knowledge) or EN ISO/IEC 17065
D1 Quality assurance of the production process	Covers both design and production. The manufacturer operates a production (manufacturing part and inspection of final product) quality assurance system in order to ensure conformity to legislative requirements (no EU-type, used as in module D without module B). The notified body assesses the production (manufacturing	

(continued)

Table 3 (continued)

Module	Description of the module	Applicable standards
	part and inspection of final product) quality system	
E Conformity to EU-type based on product quality assurance	Covers production and follows module B. The manufacturer operates a product quality (= production quality without the manufacturing part) assurance system for final product inspection and testing in order to ensure conformity to EU-type. A notified body assesses the quality system. The idea behind module E is similar to the one under module D: both are based on a quality system and follow module B. Their difference is that the quality system under module E aims to ensure the quality of the final product, while the quality system under module D (and D1 too) aims to ensure the quality of the whole production process (that includes the manufacturing part and the test of final product). E is thus similar to module D without the provisions relating to the manufacturing process	EN ISO/IEC 17021(+ product related knowledge) or EN ISO/IEC 17065
E1 Quality assurance of final product inspection and testing	Covers both design and production. The manufacturer operates a product quality (= production quality without the manufacturing part) assurance system for final product inspection and testing in order to ensure conformity to the legislative requirements (no module B (EU-type), used like E without module B). The notified body assesses the quality system. The idea behind module E1 is similar to the one under module D1: both are based on a quality system. Their difference is that the quality system under	

(continued)

Table 3 (continued)

Module	Description of the module	Applicable standards
	module E1 aims to ensure the quality of the final product, while the quality system under module D1 aims to ensure the quality of the whole production process (that includes the manufacturing part and the test of final product). E1 is thus similar to module D1 without the provisions relating to the manufacturing process	
F Conformity to EU-type based on product verification	Covers production and follows module B. The manufacturer ensures compliance of the manufactured products to approved EU-type. The notified body carries out product examinations (testing of every product or statistical checks) in order to control product conformity to EU-type. Module F is like C2 but the notified body carries out more systematic product checks	EN ISO/IEC 17025 (+ ability to decide on conformity), or EN ISO/IEC 17020, EN 17025 to be taken into account for testing required, or EN ISO/IEC 17065, EN 17025 to be taken into account for testing required
F1 Conformity based on product verification	Covers both design and production. The manufacturer ensures compliance of the manufactured products to the legislative requirements. The notified body carries out product examinations (testing of every product or statistical checks) in order to control product conformity to the legislative requirements (no EU-type, used like F without module B) Module F1 is like A2 but the notified body carries out more detailed product checks	
G Conformity based on unit verification	Covers both design and production. The manufacturer ensures compliance of the manufactured products to the legislative requirements. The	EN ISO/IEC 17020, EN 17025 to be taken into account for testing required, or EN ISO/IEC 17065, EN

(continued)

Table 3 (continued)

Module	Description of the module	Applicable standards
	notified body verifies every individual product in order to ensure conformity to legislative requirements (no EU-type)	17025 to be taken into account for testing required
H Conformity based on full quality assurance	Covers both design and production. The manufacturer operates a full quality assurance system in order to ensure conformity to legislative requirements (no EU-type). The notified body assesses the quality system	EN ISO/IEC 17021 (+ product related knowledge)
H1 Conformity based on full quality assurance plus design examination	Covers both design and production. The manufacturer operates a full quality assurance system in order to ensure conformity to legislative requirements (no EU-type). The notified body assesses the quality system and the product design and issues an EU design examination certificate. Module H1 in comparison to module H provides in addition that the notified body carries out a more detailed examination of the product design. The EU-design examination certificate must not be confused with the EU-type examination certificate of module B that attests the conformity of a specimen 'representative of the production envisaged', so that the conformity of the products may be checked against this specimen. Under EU design examination certificate of module H1, there is no such specimen. EU design examination certificate attests that the conformity of the design of the product has been checked and certified by a notified body	EN ISO/IEC 17021 (+ product related knowledge) or EN ISO/IEC 17065 or EN ISO/IEC 17020, EN 17025 to be taken into account for testing required

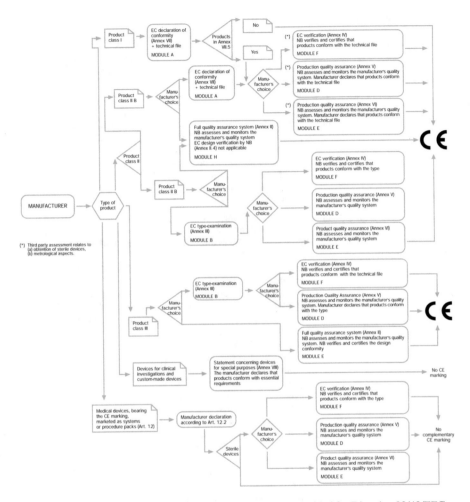

Fig. 1 Flow chart for the conformity assessment procedures provided for Directive 93/42/EEC on medical devices [18]

6 Standards Used in Manufacturing Process of Medical Devices

New Approach regulations are accompanied by a wide range of harmonised European standards. The New Approach regulatory technique is distinguished from other regulatory techniques by the relationship between regulations and standards: while the compliance with regulations is mandatory, the application of harmonised standards is in case of most New Approach directives voluntary [19].

Harmonized standards are European Norms (EN) elaborated by the European standardization bodies (CEN, CENELEC, ETSI) under a mandate from the European Commission. The Commission specifies certain essential requirements within a given directive that have to be set out in greater detail. It subsequently mandates and partially finances the development of standards that specify these details [20].

One of the most widely used standards in the area of manufacturing of the medical devices is the standard IEC 60601-1 Medical electrical equipment—Part 1: General requirements for basic safety and essential performance.

Standard IEC 60601-1 is a basic document comprised of two parts that relate to the safety of medical devices (collateral standards) and to various types of medical equipment (particular standards). The basic version of the standard was published for the first time in 1977 and was related to the issues of electrical and mechanical safety.

Standards marked with IEC 60601-1-X are collateral standards (where X represents sub-standards, altogether 11).

Standards marked with IEC 60601-2-X are specific standards (where X represents sub-standards, altogether 58).

The basic standard IEC 60601-1 is applied for the purpose of basic safety and essential performance of Medical Electronical Equipment and Medical Electrical Systems. The content of the standard describes protection against electrical Hazards from Medical Electrical Equipment, protection against mechanical Hazards of Medical Electrical Equipment and Medical Electrical Systems, protection against unwanted and excessive radiation Hazards, and protection against excessive temperatures and other Hazards.

In this part of the chapter the main activities on protection against electrical Hazards from Medical Electrical Equipment will be described, since the patients and operators are exposed to this hazard the most when operating or using the equipment which did not fulfil the requirements stated in the corresponding standard. Requirements set in the standard IEC 60601-1 give the manufacturer a possibility to better understand how to reduce risks of harm or to bring them to the acceptable limits.

Protection against electrical Hazards from Medical Electrical Equipment in accordance with standard IEC 60601 covers requirements related to maximum permissible voltage, current, energy, power sources, needed insulation, testing of leakages, etc. In order to satisfy prescribed limitation of voltage, current or energy, means for reducing the risk due to electric shock in accordance with the requirements of standard IEC 60601-1 which can be divided in two categories:

- Means of patient protection (MOPP)—Means of protection for reducing the risk of electric shock to the patient.
- Means of operator protection (MOOP)—Means of protection for reducing the risk of electric shock to persons other than the patient.

Table 4 The classification of medical devices in the USA

Class I with exemptions	General controls for devices considered as low risks for human use
Class I without exemptions	
Class II with exemptions	Performance standards for devices considered as moderate risks for human use
Class II without exemptions	
Class III	Premarket approval for devices considered as high risks for human use

Standard also covers specific measurement tests of current leakage, insulation requirements, creepage distances and air clearances.

7 Conformity Assessment in Accordance with Requirements in the USA

The relevant legislation for the Medical Devices in the USA is under the responsibility of FDA[2] (U.S. Food & Drug Administration). Medical Devices are regulated by Medical Device Amendments of 1976 to the Federal Food Drug and Cosmetic Act.

The Medical Device Amendments of 1976 to the Federal Food, Drug, and Cosmetic Administration established three regulatory classes for medical devices. These three classes are based on the degree of control necessary to assure that the various types of devices are safe and effective [21]. FDA has established classifications for approximately 1700 different generic types of devices and grouped them into 16 medical specialties referred to as panels.

The classification of medical devices in the USA is given in Table 4.

The USA has demands for marking of certain medical devices before their placement in the market, which must have Premarket Approval (PMA) equal to the European CE mark.

The most regulated devices are in Class III, as defined in the amendments as devices that support or sustain human life or are of substantial importance in preventing impairment of human health or presenting a potential, unreasonable risk of illness or injury. Insufficient information exists within Class III devices so that performance standards (Class II) or general controls (Class I) cannot provide reasonable assurance that the device is safe and effective for its intended use. Under Section 515 of the act, all devices placed into Class III are subject to premarket

---

[2]Medical Device Amendment—an amendment to the Food, Drug, and Cosmetic Act signed into law on May 28, 1976. The amendments gave FDA authority to regulate medical devices. FDA issues all approvals and regulatory required approvals depending on the class of the medical device [Marketing Clearance (510 K) or an Approval Letter (PMA)].

approval requirements. Premarket approval by FDA is the required process of scientific review to ensure the safety and effectiveness of Class III devices.

The official website of the FDA gives a wide range of information related to the medical devices which are useful for the manufacturer and other interested parties (user). Placing the medical device in the market of the USA consists of five steps [22]. Those steps are the following:

- Classify Your Device
- Select the Correct Premarket Submission
- Prepare the Appropriate Information for the Premarket Submission
- Send Your Premarket Submission to the FDA and Interact with FDA Staff during Review
- Complete Establishment Registration and Device Listing.

Premarket Notification 510(k)

Before marketing a device, each submitter must receive an order, in the form of a letter, from the FDA which finds the device to be substantially equivalent and states that the device can be marketed in the USA.

A 510(k) requires demonstration of substantial equivalence to another legally U. S. marketed device. Substantial equivalence means that the new device is at least as safe and effective as the predicate. Substantial equivalence is established with respect to intended use, design, energy used or delivered, materials, chemical composition, manufacturing process, performance, safety, effectiveness, labelling, biocompatibility, standards, and other characteristics, as applicable.

The following four categories of parties must submit a 510(k) to the FDA:

- Domestic manufacturers introducing a device to the US market;
- Specification developers introducing a device to the US market;
- Re-packers or re-labellers who make labelling changes or whose operations significantly affect the device;
- Foreign manufacturers/ exporters or US representatives of foreign manufacturers/ exporters introducing a device to the US market.

510(k) is required for anyone who wants to sell a device in the USA after May 28, 1976 (effective date of the Medical Device Amendments to the Act), if the purpose is different from the intended use for a device which is already in commercial distribution and if there is a change or modification of a legally marketed device and that change could significantly affect its safety or effectiveness.

Third Party Review

The Accredited Persons Program was created by the FDA Modernization Act of 1997 (FDAMA), based on an FDA pilot. The purpose of the program is to improve the efficiency and timeliness of FDA's 510(k) process, the process by which most medical devices receive marketing clearance in the USA. Under the program, the FDA has accredited third parties (Accredited Persons) that are authorized to conduct the primary review of 510(k) for eligible devices. Persons who are required to submit 510(k) for these devices may elect to contract with an Accredited Person and submit a 510(k) directly to the Accredited Person. The Accredited Person conducts the primary review of the 510(k), and forwards the review, recommendation, and the 510(k) to the FDA. By law, the FDA must issue a final determination within 30 days after receiving the recommendation of an Accredited Person. 510(k) submitters who do not wish to use an Accredited Person may submit their 510(k) directly to the FDA.

Once all described steps have been conducted, the medical device can be placed in the US market.

8   International Organisation for Health

At the international level there is a World Health Organization (WHO) established in 1948, (7th of April is celebrated as World Health Day). WHO represents a coordinating institution that is focused on international health within the United Nations. Under its domain, WHO realizes ensuring a leading role in the critical issues of health care and participation in partnerships where joint action is needed, preparing agendas and the transfer of knowledge, setting standards, promotes and supervises the implementation of standards, providing technical assistance in the construction of appropriate infrastructure, and controls the health status and evaluation of health trends. WHO publishes the International Health Regulations (IHR), that comprises a set of international rules structured to prevent and respond to acute public health risks, whether it is national, regional or international risk that have the potential to expand beyond the borders and threaten the health of people worldwide.

IHR represents international legal instrument that is binding for 196 countries around the world (including Bosnia and Herzegovina). In accordance with the third edition of the IHR's from 2005, which entered into force on June 15, 2007, the EC and its member states strongly support to the IHR and they will continuously support it in the future without any restrictions. IHR define the rights and obligations of countries to report public health events, and establish a number of procedures that WHO must follow in its work to uphold global public health security. Each country should have in place the infrastructure to monitor and measure the health risks with the aim to ensure the well-being of their citizens, and to prevent and control the epidemic outbreaks.

Also, on international level there is The Global Harmonization Task Force (GHTF) which was founded in 1993 by the governments and industry representatives of Australia, Canada, Japan, the European Union, and the United States of America to address the corresponding issues.

The purpose of the GHTF is to encourage a convergence in standards and regulatory practices related to the safety, performance and quality of medical devices. GHTF has done an overview of requirements for medical devices before placing it in the market by members of GHTF. From this overview it is possible to see the differences in approaches between the EU and the USA before placing the product in the market.

9 Medicine in the Field of Legal Metrology

Medical measurements are present in everyday life and they represent the basic process of prevention, diagnosis and treatment of diseases. In Europe there is an increased interest in the metrology decisions and conformity assessment decisions, which are of particular importance to carry out measurements in order to protect health.

Products with a measuring function must be designed and manufactured in such a way to provide sufficient accuracy and stability within appropriate limits of accuracy, taking into account the intention of the use of the product. The accuracy limits (permissible errors) are specified by the manufacturer himself. The measurements done by the device with a measuring function must be expressed in legal units of measurement in accordance with the provisions of Directive 80/181/EEC [23].

Metrology with its measurements is an integral part of our daily life. All measurements which are carried out with the purpose of any economic transactions, or measurements with which it's possible to take certain legal measures against or in someone's benefit, protection in the field of health and the environment, are to be classified as measurements of legal metrology. Individual governments proscribe regulations under legal metrology to meet its needs, except for the harmonized area which is equal and obligatory in all member states (11 measuring instruments, MID [24] and NAWI [25] Directive). One issue is common for all state economies and that is a fact that legal metrology is founded for the purpose of protection of the consumers (end users). OIML—International Organization of Legal Metrology has divided this category of metrology into four parts.

Those parts are:

- Legal Metrology and Trade
- Legal Metrology and Safety
- Legal Metrology and **Health**
- Legal Metrology and the Environment.

The accurate and precise measurements in the field of medicine allow easier diagnosis and identification of diseases on the basis of which it is possible to precisely determine the appropriate treatment of the patient, in order to help the patient in the best and safest way to receive effective treatment, but all through the usage of adequate medical instruments/ devices which fulfil the requirements described in the relevant legislation and standards.

In accordance with OIML D1 [26] document, governmental regulatory responsibilities include **health**, safety and environmental law. While these functions are disparate in nature, a common feature is that compliance with the law depends on measurement results. The scope of legal metrology may be different from country to country.

Therefore, the process of measurement is of direct concern to the government. Providing the laws and regulations, controlling measurement through market supervision and developing and maintaining the infrastructure that can support the accuracy of these measurements (e.g. through traceability) is essential in fulfilling the role of government.

The scope of the legal metrology regulations (e.g. which types of measurements and measuring instruments or systems are subject to legal requirements) will depend on those markets that are important to the economy, on the categories of users that the government considers necessary to protect, and on the ability of the users to protecting themselves against abuse.

Since there are only 11 harmonized instruments, it is easy to conclude that the non-harmonized sector comprises of much higher number of instruments.

Non-harmonised sectors [27] are not subject to common EU rules and may come under the national rules. These sectors should still benefit from Treaty provisions governing free movement of goods according to Arts. 34–36 TFEU. National rules on these products are subject to a notification procedure that ensures they do not create undue barriers to trade.

In order to ensure the free movement of goods in non-harmonised sectors, the principals of mutual recognition, the 98/34 notification procedure and the application of Arts 34–36 TFEU are essential.

In some regions, due to the treaties or agreements, regional legislation may have precedence over national laws and regulations or may be recommended to national authorities. This is the case for example in the European Union, where European Regulations and European Directives are accorded higher status than national legislation.

Referring again to OIML Document D1 "Considerations for a Law on Metrology", the recommendation for government bodies building their metrology systems are encouraged to keep the following:

The priority is to set up the legal provisions related to the status of the bodies to which tasks will be allocated, and the financial provisions that will ensure their sustainability (national institutes, accreditation bodies), the general framework for legal metrological control and the first list of priorities for categories to be subjected to legal control, and the infringements, penalties and the powers of agents in charge of metrological supervision.

The scope of legal metrology that is the list of categories of measuring instruments must start with the most important categories for which the available resources allow the regulation to be correctly enforced. The scope can then be progressively extended as additional resources become available.

The obligations resulting from the OIML Treaty and from the WTO TBT Agreement (obligation to use OIML Recommendations as far as possible, and encouragement by the TBT Agreement to participate in OIML recognition and acceptance arrangements) should also be taken into account, as well as other obligations deriving from regional treaties or agreements.

Measuring instruments under legal metrology have to be regularly verified. Verification [28] of a measuring instrument represents conformity assessment procedure (other than type evaluation) which results in the affixing of a verification mark and/or issuing of a verification certificate.

As described in OIML Document D1 "Fields of use of measuring instruments subject to verification" Instruments, substances, and devices used in the diagnosis and treatment of humans and animals, in the manufacture of medicines, and in the monitoring of the medical environment (patient and hospital) should be considered for verification.

OIML has published a certain number of recommendations which indicate on verification procedures of medical devices. This will be described more in detail in the chapter dedicated to legal metrology. Medical devices covered by International Organization of Legal Metrology [27] are as follows:

- Sphygmomanometers, covered by OIML recommendation R 16-1 (from 2002)
- Medical syringes, covered by OIML recommendation R 26 (1978)
- Standard graduated pipettes for verification officers, covered by OIML recommendation R 40 (1981)
- Electroencephalographs—Metrological characteristics—Methods and equipment for verification, covered by OIML recommendation R 89 (1990)
- Electrocardiographs—Metrological characteristics—Methods and equipment for verification, covered by OIML recommendation R 90 (1990)
- Measuring instrumentation for human response to vibration, covered by OIML recommendation R 103 (1992)
- Pure-tone audiometers (including Annexes A to E), covered by OIML recommendation R 104 (1993)
- Clinical electrical thermometers for continuous measurement, covered by OIML recommendation R 114 (1995)
- Clinical electrical thermometers with maximum device, covered by OIML recommendation R 115 (1995)
- Equipment for speech audiometry, covered by OIML recommendation R 122 (1996).

OIML has covered only a part of medical devices with measuring function; however the number of medical devices in use is much higher. For medical devices that are applied in the EU but which are not covered by legal metrology in some countries calibration process has to be ensured with an adequate traceability chain. Requirements for harmonizing a large number of medical devices in legal metrology are constantly increasing.

One of the countries, namely Bosnia and Herzegovina, which is not a member of the EU, but it regularly takes part in WELMEC[3] (European Cooperation in Legal Metrology) activities and OIML activities has extended the list of medical devices in the field of legal metrology in accordance with its national needs.

Medical instruments with measuring function, which are part of the Legal metrology system in B&H, are as follows:

- Defibrillator,
- Infusomats and perfusors,

- Patient monitors,
- Neonatal and paediatric incubators,
- Respirators,
- Anaesthesiology machines,
- Therapeutic ultrasound devices,
- Dialysis machines,
- Electrocardiographs ECG/EKG.

This action proved to be very successful, according to the feedback from the legal entities responsible for performing verification in this subjected field. The majority of devices tested have shown certain non-conformities related to the requirements prescribed in legal documents and/or by the manufacturer [29–33].

Unlike other measuring instruments covered by legal metrology, many of the medical devices need to be traceable to one measuring standard which is not fully developed yet. The future of development in this field of metrology lies in facilitating the process of calibration, i.e. establishment of a traceability chain via a single measuring standard for a certain type of medical device.

Turkey has also already recognised importance of metrology in medicine and has initiated via Turkish Institute of Metrology (UME) [34] a study on appropriate traceability of medical devices. The aim of the study is to develop a five year roadmap and a plan for providing reliability and metrological traceability in medical measurements. To get closer to intended aim, UME has established a Medical metrology research laboratory.

Certain measuring instruments used in medical purposes are also recognised in the Portuguese legislation [35]. The Institute of Metrology of Portugal dealt also with the question on how to improve metrological traceability of medical instruments.

Comparing the current situation of those countries dealing with medical instruments in a controlled area it is obvious that the main issue in general refers on how to ensure or improve adequate metrological traceability.

References

1. http://www.eui.eu/Research/Library/ResearchGuides/EuropeanInformation/EULegislation. aspx
2. https://europa.eu/european-union/law/legal-acts_en
3. http://eur-lex.europa.eu/homepage.html?locale=en
4. https://ec.europa.eu/growth/single-market/european-standards/harmonised-standards_en
5. The European Committee for Standardization
6. The European Committee for Electrotechnical Standardization
7. The European Telecommunications Standards Institute
8. Council Directive 93/42/EEC
9. Approval of Medical Devices, The Law Library of Congress, Global Legal Research Center
10. http://www.ema.europa.eu/ema/
11. https://ec.europa.eu/growth/single-market/european-standards/harmonised-standards/medical-devices_en

12. www.dei.gov.ba/bih_i_eu/RIA_u_BiH/?id=6634
13. http://ec.europa.eu/growth/tools-databases/nando/index.cfm?fuseaction=directive. notifiedbody&dir_id=13
14. http://ec.europa.eu/consumers/sectors/medical-devices/files/meddev/2_4_1_rev_9_classification_en.pdf
15. http://ec.europa.eu/DocsRoom/documents/18027/
16. Commission Implementing Regulation (EU) No 920/2013 of 24 September 2013 on the designation and the supervision of notified bodies under Council Directive 90/385/EEC on active implantable medical devices and Council Directive 93/42/EEC on medical devices 'designating authority' means the authority(ies) entrusted by a Member State to assess, designate, notify and monitor notified bodies under Directive 90/385/EEC or Directive 93/42/EEC
17. http://eur-lex.europa.eu/LexUriServ/LexUriServ.do?uri=OJ:L:2008:218:0030:0047:en:PDF
18. http://www.ce-marking.eu/medical-devices-class-IIb.html
19. www.dei.gov.ba/bih_i_eu/RIA_u_BiH/?id=6634
20. Stinshoff KE Role of standards in the assessment of medical devices. https://www.iso.org/files/live/sites/isoorg/files/archive/pdf/en/wsc-medtech_10_klaus_stinshoff_text.pdf
21. http://www.fda.gov/medicaldevices/productsandmedicalprocedures/deviceapprovalsandclearances/pmaapprovals/default.htm
22. http://www.fda.gov/MedicalDevices/DeviceRegulationandGuidance/HowtoMarketYourDevice/default.htm
23. Council Directive of December 20, 1979 on the approximation of the laws of the Member States relating to units of measurement and on the repeal of Directive 71/354/EEC
24. Directive 2014/32/EU of The European Parliament and of The Council of February 26, 2014 on the harmonisation of the laws of the Member States relating to the making available on the market of measuring instruments
25. Directive 2014/31/EU of The European Parliament and of The Council of February 26, 2014 on the harmonisation of the laws of the Member States relating to the making available on the market of non-automatic weighing instruments
26. OIML D 1 Considerations for a Law on Metrology
27. https://ec.europa.eu/growth/single-market/goods/free-movement-sectors_en
28. International vocabulary of terms in legal metrology (VIML)
29. Badnjevic A, Gurbeta L, Jimenez ER, Iadanza E (2017) Testing of mechanical ventilators and infant incubators in healthcare institutions. Technol Health Care J 25:237
30. Gurbeta L, Badnjevic A, Dzemic Z, Jimenez ER, Jakupovic A (2016) Testing of therapeutic ultrasound in healthcare institutions. In: Bosnia and Herzegovina, 2nd EAI international conference on future access enablers of ubiquitous and intelligent infrastructures, 24–25 October 2016, Belgrade, Serbia
31. Gurbeta L, Badnjevic A, Pinjo N, Ljumic F (2015) Software package for tracking status of inspection dates and reports of medical devices in healthcare institutions of Bosnia and Herzegovina. In: XXV international conference on information, communication and automation technologies (IEEE ICAT), pp 1–5, 29–31 October 2015, Sarajevo, Bosnia and Herzegovina
32. Gurbeta L, Badnjevic A, Sejdinovic D, Alic B, Abd El-Ilah L, Zunic E (2016) Software solution for tracking inspection processes of medical devices from legal metrology system. In: XIV mediterranean conference on medical and biological engineering and computing (MEDICON), 31. March–2 April 2016, Paphos, Cyprus
33. Gurbeta L, Badnjevic A (2016) Inspection process of medical devices in healthcare institutions: software solution. Health Technol. doi:10.1007/s12553-016-0154-2
34. Karaböce B, Gülmez Y, Akgöz M, Kaykısızlı H, Yalçınkaya B, Dorosinskiy L (2015) Medical metrology studies at Tübitak UME. 17 international congress of metrology, 0 011
35. The role of metrology in medical devices MARIA DO CÉU FERREIRA Instituto Português da Qualidade, Portugal, OIML Bulletin Volume LII Number 4 October 2011

Legal Metrology System—Past, Present, Future

Alen Bošnjaković, Almir Badnjević and Zijad Džemić

Abstract Through the chapter Legal Metrology System—Past, Present, Future a significant role of legal metrology is presented in the area of development of human civilization, which has led to the consistency of measurements in everyday life. The current state of legal metrology on the International and on the European level is described. Furthermore, through the chapter the legislation of the European Union is presented, which includes directives of the New Approach, with special reference to the Medical Devices Directive, which deals with medical devices with a measuring function. Through the chapter the necessity, as well as the challenges for placing these devices into the framework of legal metrology is presented. In addition, the role of legal metrology and the factors that will have an impact on its development in the future is described.

1 Introduction

Metrology (derived from the Greek word "metron" meaning measure, and logos meaning science, studying) represents a basis for development of all scientific disciplines including medicine and the usage of medical devices with measuring function. Metrology development is related to the development of human civilization, which required consistency of measurements in everyday life. These requirements have shaped a uniform measuring system that eventually grew into the SI system.

A. Bošnjaković (✉) · Z. Džemić
Institute of Metrology of Bosnia and Herzegovina (IMBIH), Sarajevo,
Bosnia and Herzegovina
e-mail: alen.bosnjakovic@met.gov.ba

A. Badnjević
Medical Devices Verification Laboratory, Verlab Ltd Sarajevo,
Sarajevo, Bosnia and Herzegovina

© Springer Nature Singapore Pte Ltd. 2018
A. Badnjević et al. (eds.), *Inspection of Medical Devices*, Series in Biomedical
Engineering, https://doi.org/10.1007/978-981-10-6650-4_3

Metrology, as the science of measurement, embraces both experimental and theoretical determinations at any level of uncertainty in any field of science and technology [1].

Metrology primarily deals with:

- Measuring units,
- Development, realization, maintenance, and improvement of standards (SI), dissemination of the values reproduced to achieve traceability,
- Measuring instruments,
- Measuring, their uncertainty, and reliability.

Generally, Metrology is divided into three main areas covering different levels and accuracy:

- **Scientific metrology**—represents the highest level of metrology and deals with organization and development of measurement standards and their maintenance,
- **Industrial metrology**—deals with the measuring instruments used in industry, in production processes and testing procedures, which ensures quality of products when properly functioning, as well as competitiveness and customer satisfaction, and
- **Legal metrology**—deals with ensuring the accuracy of results for measuring instruments that has an impact on transparency of transactions, security, environmental protection, and **health protection**.

Fundamental Metrology has no internationally accepted definition. It includes studying the general terms of the highest level of accuracy of measurements in certain areas. Fundamental Metrology can be described as a part of the advanced level of scientific metrology.

Scientific Metrology makes scientific researchers improve the definitions and accuracy of SI units. It deals with standard organization, development, and maintenance that help in realizing primary values of SI units, develops and improves calibration methods, and disseminations of the units from high-level standard to lower ranked standard.

Legal metrology deals with measuring instruments and measurements when it is necessary to ensure essential fair trade, security, environmental and health protection. This is achieved by an adequate system of legal metrology, or by the control of measuring instruments.

Legal metrology is regulated by laws and regulations, and it can vary from country to country. Authorized bodies for legal metrology control the legal metrology system according to the needs and opportunities of a particular country. Insurance of the metrology system includes all regulations, technical tools, and necessary actions that can be used to secure measurement results' credibility in legal metrology.

Measurements are one of the first technical activities in which international cooperation is established. Internationalization and harmonization are essential features of measurement.

The basis for dealing with measurement issues at an international level are:

- **Metre Convention, and**
- **International Organization of Legal Metrology (OIML)**

Metre Convention is an international agreement which is established by the International Bureau of Weights and Measures (Bureau International des Poids et Mesures—BIPM). This diplomatic contraction was signed by representatives of 17 countries in Paris in 1875. Also, the structure was established, which helped governments of Member States to deal with metrology issues [2].

International Convention for establishing the International Organization of Legal Metrology (Organisation Internationale de Métrologie Légale—OIML) was signed in 1955, and it represented intergovernmental organization that was formed with the aim to perform activities that are not included in scientific metrology within the Metre Convention [3].

The legal control of measuring instruments, which are used in the area of legal metrology, includes the following activities [4]:

- Type evaluation and approval:
  - testing laboratories,
  - certification bodies (issuing authorities).

- Initial verification:
  - field officials,
  - manufacturer's declaration.

- Subsequent verification:
  - field officials,
  - readjustment (calibration),
  - maintenance and repair.

- Market surveillance:
  - individual instrument failures identified, recorded and notified,
  - recall of instrument types displaying a record of failures,
  - requires manufacturers to implement adjustments in the area or in production.

Primary terms and definitions that are used in the legal metrology field are given in the text below [5, 6]:

Legal metrology Practice and process of applying statutory and regulatory structure and enforcement to metrology

Note 1 The scope of legal metrology may be different from country to country.

Note 2 Legal metrology includes:

- setting up legal requirements,
- control/conformity assessment of regulated products and regulated activities,
- supervision of regulated products and of regulated activities, and
- provision of necessary infrastructure for the traceability of regulated measurements and measuring instruments to SI units or national standards.

Note 3 There are also regulations outside the area of legal metrology pertaining to the accuracy and correctness of measurement methods.

Type (pattern) evaluation A conformity assessment procedure must be done on one or more specimens of an identified type (pattern) of measuring instruments, which results in an evaluation report and/or an evaluation certificate.

Type approval Decision of legal relevance, based on the review of the type of evaluation report that the type of a measuring instrument complies with the relevant statutory requirements and results in the issuance of the type of approval certificate.

Initial verification Verification of a measuring instrument which has not been verified previously.

Subsequent verification Verification of a measuring instrument after a previous verification.

Note 1 Subsequent verification includes:

- mandatory periodic verification,
- verification after repair, and
- voluntary verification.

Note 2 Subsequent verification of a measuring instrument may be carried out before expiry of the period of validity of the previous verification either at the request of the user (owner) or when its verification is declared as no longer valid.

Metrological supervision Activity of legal metrological control performed to check the observance of metrology laws and regulations.

Note 1 Metrological supervision also includes checking the correctness of quantities indicated on and contained in pre-packages.

Note 2 The means and methods such as market surveillance and quality management may be utilized to achieve stated application.

2 Legal Metrology System—Past, Present

The presentation of the origin of legal metrology and an overview of the current status of legal metrology at international and European level will be given in this part of the text. Also, some other regional metrology organizations responsible for this area will be shortly discussed.

Legal metrology evolved parallel to the civilization the required consistency of the broad range of measurements used in everyday life. The relationship between the country and legal metrology must be very close. Each country needs the information provided by legal metrology to be able to protect its users providing them with uniform measurements. On the other hand, legal metrology requires country authority to ensure uniformity of measurements [7].

The first step for the development of legal metrology is set through signing the International Convention for establishing the International Organization of Legal Metrology. However, the contract did not contain the legal metrology definition, so the first document containing the definition of legal metrology would be the OIML (OIML D1) document [8].

Through history, the role of the country in legal metrology is reflected through the adoption of regulations governing legal metrology area, with the aim to protect measurement users. Monitoring the legal metrology area helps creating confidence in conducted activities. On the other hand, legal metrology requires complete independence in measuring process, making it one of the main reasons for linking legal metrology with the country, i.e. state economy. A stable country implies the existence of an active legal metrology system and higher confidence in conducted measurements. Less developed countries normally have a less developed legal metrology system, but not necessarily the lower quality of performed measurements.

Each country has developed its legal metrology system through history, which contributed to the fragmentation of the market, i.e. a different approach to implementation of the legal metrology system. It is certain that such a fragmented system inevitably led to trade barriers by various requirements for measuring instruments and methods that were applied to measurements, as well as measuring units.

Changes that took place within agriculture, industry, urbanization and trade in the last three hundred years, has had the conversion of the local market into the national market and domestic market into a global market as a consequence. These changes inevitably influenced legal metrology leading to the reorganizations of this area.

Activities, role, and tasks of the International Organization of Legal Metrology (OIML) together with the bodies existing within this organization will be described in the following text. European Organization of Legal Metrology (WELMEC) and its role in legal metrology at European level will also be presented. The framework of legal metrology within the European Union will be given with medical devices particularly outlined, as well as the importance of tackling these devices under legal metrology from the aspect of user protection. Other related regional organizations will also be presented, as well as the ideas of possible future development of legal metrology.

3 International Organization of Legal Metrology—OIML

The International Organization of Legal Metrology OIML [9] was established in 1955 as the contractual intergovernmental organization. Its purpose is to enable economies creating effective legal metrology infrastructures that are mutually equal and internationally recognized, for all areas where governments have their responsibility, such as incentives by trade, the establishment of mutual confidence and to harmonize the level of costumers' protection worldwide [10]. According to the stated, the organization has given a definition of legal metrology on its official website and as it follows [11]:

> Legal metrology is the application of legal requirements to measurements and measuring instruments

Legal metrology has extensive use in trade, health, safety and environmental protection. These mentioned areas or activities of legal metrology extensively provide protection to individuals within society as well as to society in general with its legal metrology regulations. Regulations in legal metrology are related to measurement results and legal control which is performed on behalf of the government.

OIML has a goal to promote the procedures and recommendation in the field of legal metrology, as well as their global harmonization. OIML has developed structure, which allows its Members to participate in the work of OIML Technical Committee and to suggest topics that are relevant to them or for the region. OIML recommendation and documents that are related to the manufacturer and use of measuring instruments are available to OIML Members, thus avoiding unnecessary guides and documents that refer to the needs of only one country. Also, the participation in the work of OIML allows its Members to have access to the latest information from other Members regarding new technologies, type of conformity assessment, solutions of other countries in dealing with certain requirements, good practices in legal metrology, legal metrology organizations in different countries, etc. Furthermore, members of OIML can affect the OIML policy and suggest recommendations for OIML strategy. In this way, members ensure that their needs will be considered in the work of the organization. Strategy deals with general organization policy, as well as with the support of national authorized bodies of legal metrology given to its members, especially developing members. More about the benefits, which are provided by membership in the OIML, can be found on the official website [12]. According to the research, published by World Bank in 2007, members of OIML cover 86% of the world population and 96% of the global economy [12]. From the mentioned information it can be concluded that the impact of legal metrology is crucial to our daily life and that it creates huge profit from invested money.

OIML is constituted of 61 Member States and 61 Correspondence Member States (Observers) with headquarters in Paris by 2016 [13]. OIML had 57 Member States and 56 Correspondence Member States (Observers) by 2010 [14]. Based on this data, it can be concluded that there is continual interest for becoming a part of OIML community with the aim to strengthen the individual system of legal metrology as well as society protection within Member State.

Member States are those who have ratified OIML Agreement Convention and requested access to OIML membership to the French government. All Member States have morally committed that they will apply legal metrology as far as possible in their countries.

Correspondent Members (Observers) are countries or economies that cannot become, or have not yet become the full member but are interested in the work of OIML and want to participate in it. These countries may be represented by an authorized government body, authorized legal metrology body, or authorized trade chamber or may be represented by state institutes or institutes responsible for legal metrology issues in that country.

International Organization of Legal Metrology OIML has the following tasks:

- It develops regulations, standards and similar documents used by authorized metrology bodies and industry,
- It provides a system of mutual recognition which reduces trade barriers as well as costs on the global market,
- It represents interests of the legal metrology community in the frame of international organizations and forums dealing with metrology, standardization, certification, testing and accreditation,
- It promotes and facilitates transfer of knowledge and competences of legal metrology community worldwide, and
- It cooperates with other authorities to raise awareness about importance of the contribution of uniform legal metrology infrastructure, which can contribute to the development of a modern economy.

International Organization of Legal Metrology OIML has the following authorities [15]:

- International Conference of Legal Metrology,
- International Committee of Legal Metrology,
- Technical Committees (TC),
- International Bureau of Legal Metrology.

International Conference of Legal Metrology is the supreme body of the Organization. It consists of representatives of the Member States always coming from an authorized body of legal metrology. Also, Correspondence Members (Observers) have the possibility to access the Conference as Observers and organizations that OIML cooperates with may be invited to attend the meeting as Observers. The conference meets every four years. The conference's role and its field of action is the following [16].

- It studies issues related to the goals of OIML and based on their making decisions,

- It provides the structure of Administrative bodies,
- It confirms the members of Technical Committee,
- It examines and approves reports submitted by the bodies of OIML, and
- It makes adoption and approval of the budget and determination of the amount of registration fee for OIML members.

International Committee of Legal Metrology (CIML) [17] is a working body in the organization of OIML. CIML. Members (one per country) are designated by the government of the country of origin. Designated members are usually officers in departments that deal with measuring instruments or play the official active role in the field of legal metrology. CIML members are permanent OIML partners and in the Member States, they have the double role: to represent their country at the Conference and to represent OIML in their country. Correspondence members (Observers) have the possibility to access CIML as Observers, and organizations which CIML cooperates with, may be invited to attend the meeting as Observers. The Committee meets once a year, and in its work scope brings the following resolutions [18]:

- It approves of OIML strategy and annual working plan BIML [15],
- It approves of the financial report of the director of BIML,
- It designates the BIML Director and Assistant Director,
- It adopts different international regulations, procedures, etc.,
- It approves of amendments related to the technical program of the OIML, and
- It adopts the OIML recommendations, documents and other publications, etc.

Technical Committees (TC) consists of representatives from the Member States as well as Correspondence Members (Observers) and organizations that cooperate with OIML. Committees are responsible for making international recommendations and documents based on tasks assigned by CIML, that provide a legal basis for the adoption of legislation for different types of measuring instruments in the scope of legal metrology of corresponding Member States. Only Member States representatives have the right to vote in the committees. Altogether there are 18 Technical Committees, and 50 Subcommittees, whose tasks include:

- The scope and application of metrology,
- Metrological requirements,
- Methods and equipment for testing and verification of compliance with the metrological requirements, etc.

Technical Committees (TC) which work and operate under the supervision of the CIML are [19]:

- TC 1 Terminology;
- TC 2 Units of measurement;
- TC 3 Metrological control;
- TC 4 Measurement standards and calibration and verification devices;
- TC 5 General requirements for measuring instruments;

- TC 6 Pre-packaged products;
- TC 7 Measuring instruments for length and associated quantities;
- TC 8 Measurement of quantities of fluids;
- TC 9 Instruments for measuring mass and density;
- TC 10 Instruments for measuring pressure, force and associated quantities;
- TC 11 Instruments for measuring temperature and associated quantities;
- TC 12 Instruments for measuring electrical quantities;
- TC 13 Measuring instruments for acoustics and vibration;
- TC 14 Measuring instruments used for optics;
- TC 15 Measuring instruments for ionizing radiations;
- TC 16 Instruments for measuring pollutants;
- TC 17 Instruments for physic-chemical measurements;
- TC 18 Medical measuring instruments.

International Bureau of Legal Metrology (BIML) is the executive body of OIML, which ensures that the activities and long-term planned activities proceed according to the plan. BIML responsibilities are the following [20]:

- It organizes meetings of the Conference and of the Committee,
- It takes care of the implementation of decisions that have been made in Conferences and Committees,
- It prepares the Conference and meetings of Technical Committees,
- It supervises and coordinates the work of the OIML Technical Committees, publishes and distributes publications of the organization,
- It publishes an OIML newsletter, and takes care of the official site,
- It maintains contacts with organizations and performs activities related to the organization, and
- It prepares and implements the budget of the organization, and shows the CIMLs financial report.

The primary activity of different technical committees (TC) is to develop model regulations as well as international recommendations. Recommendations provide the Member States, as well as Contracted Members (Observers) with the basis for the establishment of unified national legislation that reduces trade barriers, but also provides protection for manufacturers and users of measuring instruments [4]. OIML is an Observer in the WTO (TBT) [21] Committee, and international OIML recommendations represent international standards regarding the Agreement on technical recommendations in trade [22]. Considering the broad application of international recommendations, more and more manufacturers decide to implement these recommendations, with the aim of the fulfilment of prescribed measuring instruments features. Recommendations are mostly related to the type of approval and contain the following recommendations for type of approval and certification [4]:

1. Metrological requirements:

   - Accuracy class,
   - Maximum permissible error:

     - rated operating conditions, reference conditions,
     - rated operating conditions with influential factors;

   - Influential factors:

     - ambient conditions (temperature, humidity, etc.),
     - mechanical factors,
     - electromagnetic factor

   - Repeatability and reproducibility,
   - Discrimination and sensitivity,
   - Reliability over time, and
   - Mutual recognition and acceptances arrangements

2. Technical requirements:

   - Indication of the results,
   - Software,
   - Markings,
   - Operating instruction, and
   - Suitability for use

3. Test program and procedures,
4. Format and report tests,
5. Conformity declaration or certification.

Furthermore, OIML recommendations are related to verification, because initial and subsequent verification are also activities of legal metrology according to the national or regional regulations.

OIML system [23] was established in 1991 with the purpose to simplify administrative procedures and to reduce costs related to international trade of measuring instruments in the field of legal metrology. In the beginning, the system was called the OIML Certificate System, and now it is called the OIML Basic Certificate of Conformity. The purpose is clearly to make the difference between OIML Basic Certificate of Conformity and the OIML MAA Certificate System. This system provides the opportunity for the manufacturer to get an OIML Basic Certificate and an OIML Basic Report (Test report) based on which is confirmed that the given instrument complies with the requirements of the type, that are prescribed by relevant OIML international recommendation. All information related to the system, certain steps, rules, and conditions for use, edition, and use of OIML certificates, could be found in the OIML document OIML B3 [24]. This certificate is issued by OIML Member States, which may happen to have one or more bodies for instruments testing [25]. This type of certificate can be accepted by other countries without any additional examinations. This helps manufacturers place their products on the global market.

OIML MAA refers to type testing, and its purpose is to increase the level of mutual confidence that is given by the OIML Basic Certificate. The OIML MAA is voluntary, but it provides a lot of benefits. It enhances the level of trust in conducted assessment for laboratory testing, which is included in type testing. It brings benefits to OIML Member States and Correspondent Members (Observers), which do not have established laboratories for testing. It gives the opportunity to have considered (Declaration of Mutual Confidence—DoMC) the additional state requirements or regional requirements (which can only apply relevant OIML recommendations). The purpose of MAA for the participants is to accept and use MAA test reports, which are verified by the OIML MAA Certificate of Conformity. Participants in the MAA either publish (Issuing Participants) or use (Utilizing Participants) test reports. The benefit for manufacturers is to avoid duplication of tests in different countries. The participants confirm their participation in the system by signing the Declaration of Mutual recognition. Implementation of the OIML MAA started in 2005 [25]. All information related to the system, certain steps, rules, and conditions for use, edition, and the use of OIML certificates, can be found in the OIML document OIML B10 [26].

As for the OIML Basic Certificate published by authorized bodies, it has to demonstrate compliance with the requirements according to ISO/IEC 17065 [27] by using the results from laboratories for testing which have implemented the requirements according to ISO/IEC 17025 [28]. The OIML MAA confidence in test reports is followed by an official and mandatory peer review process. This process proves the compliance of authorized bodies for certification publishing (OIML Member States) with testing laboratories applying required standards and also, the possibility of labs for testing to perform tests. In order to prove their competence, authorized bodies for certification publishing and laboratories for testing must be accredited for the area including the OIML recommendations or they must pass through the peer review process [29], respectively.

Global legal metrology perspective is located within this Organization, and it reflects through the harmonization process of legal metrology, as well as trade and administrative barriers removal with a uniform approach in this area. Considering that OIML provides infrastructural support for developing countries, countries that still did not achieve OIML membership have a motivation to get involved in the largest legal metrology organization in the world with the purpose to improve their legal metrology system. As for developed countries, their motive would be constant improvement of their metrology system through exchange experience with other countries.

4 European Legal Metrology—WELMEC

Legal metrology consists of technical and administrative procedures, established by law and by authorized bodies with the purpose to guarantee the quality of measurements performed during the commercial transaction, official controls of certain health care, safety, etc.

WELMEC (Western European Cooperation in Legal Metrology) was established 1990 by signing the Memorandum of Understanding among 18 [30] Member States, which represented authorized bodies of legal metrology in the European Union, and three EFTA [31] Member States.

WELMEC Committee is made of three separate categories, and those are Members, Associate Members, and Observers. At the beginning of the organization establishment, there were altogether 18 Members and 10 Associated Members. The Associated Members come from Central and Eastern Europe countries.

Currently, WELMEC Members are national bodies for legal metrology in the Member States of European Union and EFTA, i.e. altogether it contains 31 Members and 6 Associate Members. WELMECs primary purpose is to establish a harmonized structure in application of European Legal Metrology as well as [30]:

- It develops confidence among national bodies of legal metrology,
- It harmonizes the activities regarding legal metrology,
- It identifies specific features of legal metrology, which should be reflected in frames of European metrology, certification, and testing,
- It enables exchange information among the Member States,
- It promotes legal metrology for the purpose of removing trade barriers in the field of measuring instruments,
- It promotes consistency in interpretation and application of regulatory documents and activities suggestions to facilitate their implementation,
- It identifies technical problems, which may happen to be subject to cooperation,
- It maintains a working relationship with other relevant bodies with interest in legal metrology,
- It supports discussion about trends and establishing criteria for legal metrology and to maintain channels for continued transfer knowledge.

WELMEC plays a consultative role in the European Commission and European Union Council related to the implementation, and further development of legal metrology in the European Union. WELMEC Committee consists of Members and Observers delegates and other regional bodies that have an interest in legal metrology. The Committee meets once a year, and its work is supported by working groups, and those working groups will be explained in further detail in the next section. Also, organization working groups create guidelines to harmonize and simplify implementation of a legal metrology in the European Union, while, at the same time, ensuring protection of the end user of services in legal metrology, as well as of the manufacturer (Fig. 1).

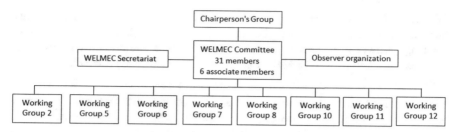

Fig. 1 Organization scheme of WELMEC [32]

Working Group 2 (Non-automatic and Automatic Weighing Instruments) works on implementation of Directive for **Non-automatic Weighing instruments— NAWI (2014/31/EU)** [33] and **Measuring Instruments Directive MID (2014/32/EC)** [34] in regard to Automatic weighing instruments (**AWIs**). The purpose of considering these Directives is reflected in aspects of implementation and not only in control, entry in into force and subsequent verification. Also, the working group, while working uses the following OIML Recommendation: OIML R 47—Standard weights for testing of high capacity weighing machines, OIML R52—Hexagonal weights—Metrological and technical requirements, OIML R111—(all parts) "Weights of classes E_1, E_2, F_1, F_2, M_1, M_{1-2}, M_2, M_{2-3} and M_3", and OIML 134—(all parts) "Automatic instruments for weighing road vehicles in motion and measuring axle loads".

Working Group 5 (Measurement Surveillance) works on establishing confidence in the field of legal metrology by cooperating with the WELMEC Member States with the aim of promoting equally valuable metrological control level in the European Union. Also, it works on establishing information exchange and guidelines for measurement surveillance including market surveillance, and inspection among Members of WELMEC.

Working Group 6 (Pre-packed products) works on promotion of equal Directive and regulation applications, which are related to these products (Pre-packed products, nominal quantity, packaging, and sampling). Also, the group works on establishing information through the exchange platform, which handles each question that refers to pre-packed products, which are sold in a quantity that is given either in European or National legislation. The group offers European Commission advice about aspects that need to be considered in new and existing regulation. The group works to develop a guide for pre-packed products.

Working Group 7 (Software) works to harmonize the process of type approval and equivalent software conformity assessment procedures, which are necessary for measurement instrument features. These activities are carried out through the development of specific guidelines, as well as through considering of technologies that occur in the use of measuring instruments, i.e. with Measuring Instruments Directive and Non-automatic Weighing Instruments Directives. Also, this group uses OIML Recommendations in its work.

Working Group 8 (Measuring Instrument Directive) deals with general application of Measuring Instruments Directive (MID), as well as with information exchange between WELMEC Members. Besides the already stated, this group considers definition of WELMEC and OIML and provides explanation, where it happens to spot differences.

Working Group 10 (Measuring equipment for liquids other than water) works on harmonization of a procedure for type approval tests and first verification for non-water fluids, by developing guidelines. The Working group uses Measuring Instruments Directive (MID) and OIML Recommendation in its work, such as OIML R 117—Dynamic measuring systems for liquids other than water, OIML R80—Road and rail tankers with level gauging, OIML R81—Dynamic measuring devices and systems for cryogenic liquids, etc.

Working Group 11 (Utility Meters) works on harmonization of legal metrology, related to utility meters, by implementing Measuring Instruments Directive (MID), provides support to the new measurement techniques for measuring instruments covered by Directive, the development of proposals concerning the scope and application of the Directive.

Working Group 12 (Taximeters) considers Measuring Instruments Directive (MID) and other regulatory documents or harmonized standards related to taxi meters, and it examines other aspects that refer to installing taxi meters in vehicles (signal, additional devices, etc.), not only in inspection, application, and subsequent verification.

More about working groups and work of WELMEC organization can be found on their official website http://www.welmec.org.

5   Other Regional Legal Metrology Organizations

Besides International Organization of Legal Metrology OIML and European Regional Organization of Legal Metrology WELMEC, there are also other regional organizations of legal metrology. Regional organizations of legal metrology aim to accomplish information exchange among Member States and their economies and to achieve harmonization and consistent approach to legal metrology in their region. Regional organizations in the world that deal with legal metrology are [35]:

- Inter-American Metrology System—Legal Metrology Working Group (SIM–LMWG) [36],
- Asia-Pacific Legal Metrology Forum (APLMF) [37],
- Euro-Asian Cooperation of National Metrological Institutions (COOMET) [38], and
- Euro-Mediterranean Legal Metrology Forum (EMLMF),

SIM is the result of a signed contract among state metrology organizations. SIM is currently made of 34 Member States from the Organization of American States (OAS). SIM aims to promote international cooperation and interoperability between the Member States and regional cooperation in legal metrology. SIM is dedicated to implementing a World Measuring System, in which users would have confidence, in the framework of America.

APLMF consists of a group of authorized bodies in the field of legal metrology in an area with the goal to develop legal metrology and promote free and open trade in the region, trough harmonization as well as removal of technical or administrative barriers when it comes to the trade field of legal metrology.

COOMET is an organization of national metrology institutes of Central and Eastern Europe. It was established in 1991, and in 2000 it was renamed "Euro-Asian Cooperation of National Metrological Institutions" and by this time had 21 Member States.

WELMEC provides the possibility for access to countries that are not a part of the European Union, but are located in Europe, to promote its system of legal metrology and harmonize it with legislation of Member States of the European Union. In this way, a unique economic area has been created and harmonization of requirements interpretation, which arises from the legal metrology field and which gives these economies and societies potential for their development.

6 Legislation Framework of European Union

Until the nineties, European countries had individual national legislations for measuring instruments, and they published type approvals for measuring instruments before they were put on the market (Old Approach). Entry into force **Directive for Non-automatic Weighing Instruments 90/384/EEC (NAWI)** that dates from 1993, the New approach is started and with **Directive for Measuring Instruments 2004/22/EU (MID)**, that dates from 2006, the harmonization process of legal metrology in European Union is continued (New Approach).

The European Union Legislation is based on a dual approach: Scheme approach based on directives of New and Global Approach, and Sectoral Approach. New Approach has a significant role in supporting free movement in the international European Union market. Also, New Approach limits county intervention for important issues and to the business world it gives the opportunity to choose the way on how to fulfil its obligation to the society.

The New Approach implies the fulfilment of requirements prescribed by directives for measuring instruments before putting them on the market. Fulfilling these requirements must be done by a manufacturer of measuring instruments or a Notified Body. The product (measuring instrument) after the conformity assessment procedure is marked with "CE" mark [39] and it must be accepted by the European Union Member Countries. Not fulfilling requirements for the product that is placed in the market of European Union brings sanctions for the manufacturer according to the provisions of the Member Country, in which the product is placed on the market. Both mentioned directive for measuring instruments were republished. CE mark symbolizes that the product is harmonized with all requirements from the New Approach, which applies to it.

Technical barriers in the area of legal metrology have been removed by entry into force of above mentioned directives, and the conditions for the creation of a unique economic space in the European Union were created.

Directive NAWI regulates use of measuring instruments from retail to industrial weigh measuring instruments in commercial, legal, and for medical purpose. **Directive MID** regulates use of the following measuring instruments: water meters, gas meters, electricity meters, heat meters, fuel meters, automatic weighing instruments, taximeters, materialized measures, dimensional measuring instruments, and exhaust gas analysers (in total ten types of measuring instruments).

Although directives prescribe provision for measuring instruments from legal metrology, the area of legal metrology does not cover the same scope in all European Union Member Countries. Member countries have the possibility to decide which type of measuring instruments they will prescribe metrological requirements based on their legal control. The information about measuring instruments from legal metrology for individual European Union Member Country [40], i.e. Member Country and Associate Members can be found on the official website of WELMEC.

Furthermore, with the appearance of directives, authorized national bodies designated competent bodies for performing conformity assessment of this instrument type, for publishing EC Type Approval in order to put measuring instruments on the European Union market. Manufacturers of measuring instruments from the European Union, which get EC Type Approval and EC Initial Verification based on mentioned directives, can put their measuring instruments on European Union market without re-examination and re-approval. The unique economic area is created in this way. This means that the manufacturer according to these NAWI and MID directives has an obligation to put the "CE" mark on the measuring instrument, as well as to put additional measurement requirements, and the number of the Notified Body on the instrument. By other directives, competitive bodies that are designated by the authorized national body are reported to the European Commission. Each Notified Body has its number, which allows its identification. This number is located on the measuring instrument, as mentioned above, and it is used to identify the body, which has done the testing. According to the rules specified in directives, the opportunity has been given to the manufacturer to perform verification itself, after the Notified Body has done the assessment of their quality system. In this case, the manufacturer is obliged to use the mark of the Notified Body, and to put it on measuring instruments [41]. The Notified Bodies prove their competence through a accreditation process [42] or self-declaration (that is a case by National Metrology Institutes) fulfilling criteria for acquiring the status of Notified Body. The status of Notified Bodies in the European Union can be checked in NANDO [43] database.

Currently, there are 117 Notified Bodies for MID Directive, 97 Notified Bodies for NAWI Directive, and 58 Notified Bodies for Medical Device Directive (93/42/EEC) in total. Medical Devices Directive (93/42/EEC) handles medical instruments that have measuring functions. Although it is not recognized as part of official legal metrology in the European Union, individual countries have already recognized its importance from the aspect of health, and they are gradually introducing these medical measuring instruments in the legal framework. It is expected that the largest number of Notified Bodies is in the MID directive because it contains the most significant number of devices for which is necessary for a conformity assessment to be done.

Conformity assessment implies the testing and evaluation of instruments used for determination if the single instrument, as well as the group of instruments or produced series of instruments, is harmonized with a legal basis applicable to this type of instruments [44]. Conformity assessment of instruments together with requirements of MID Directive is done according to one of the conformity assessment procedure which is chosen by the manufacturer. Conformity assessment procedures are described in annexes of Measuring Instrument Directive.

Type approval represents a decision with legal relevance, based on the review on the type of evaluation report, that the type complies with the relevant statutory requirements, and that it is suitable for use in prescribed areas in the way where reliable measurement results are expected in the prescribed time interval [45].

Measurement instruments verification is a conformity assessment procedure (other than type approval) which results in the affixing of verification mark and/or issuing of a verification certificate, which proves and confirms that the instrument is in compliance with legal requirements [45].

Entry into force of Measuring Instruments Directive (MID) in 2006 has as consequence the legislation amends of European Union Member Countries (metrology legislation) regarding the implementation of this directive and fulfilling the requirements for putting these instruments on the European Union market, respectively. MID Directive had a much bigger impact than NAWI Directive in 1993 related to the designation process of the bodies conducting the conformity assessment.

This impact results from the fact that MID Directive covers ten types of measuring instruments, and NAWI Directive covers only measuring instruments for mass (Non-automatic Weighing Instruments). This means that the approach to conformity assessment of ten new measuring instruments, which also included sub-assemblies such as converters, must be harmonized. Furthermore, the number of Notified Bodies for both directives has increased within time, because of the work scope of conformity assessment. In the conformity assessment process, Notified Bodies were not the only ones who have participated, private and commercial bodies, which demonstrated their competence to performing this type of action, participated too.

The European Parliament Decision (768/2008/EC) [41] from 2008 established the procedure for placing measuring instruments covered by NAWI and MID Directives on the market, i.e. it has set up the procedure for conformity assessment of these instruments. Also, the decision contains obligation of the European Commission to allow the information exchange between authorities responsible for national policies of designation, as well as to ensure coordination and cooperation between Notified Bodies according to existing laws or other legislation of Commission, and to act as a group of Notified Bodies [41].

According to the mentioned Decision NMi [46] (the Netherlands) took the initiative to establish a European platform for coordination between activities of Notified Bodies, which act in the legal metrology area. Notified Bodies of CMI [47] (Czech Republic), LNE [48] (France), NMO [49] (UK), METAS [50] (Switzerland), and NMI (the Netherlands) contacted other Notified Bodies in 2010 to establish the European Platform for these bodies. In this way, the platform for cooperation between Notified Bodies in EU (NoBoMet) has been formed. **NoBoMet** is a European platform for Notified Bodies, which acts in the legal metrology area and its purpose is conformity assessment of measuring instruments [51].

The purpose of the platform is to optimize the circumstances and conditions under Notified Bodies work and to make the platform transparent for manufacturers of measuring instruments for which conformity assessments are prescribed. The platform has a goal to encourage the performance equality of Notified Bodies regardless of the circumstances. The purpose of this platform is to be recognized as the representative of Notified Bodies in Europe and to ensure consistent legal metrology infrastructure. Similarly, this platform has the aim to be interlocutor to internationally recognized committees such as the EU, WELMEC, and OIML. The Notified Bodies apply OIML recommendations as normative documents for the purpose of conformity assessment [41]. Currently, Notified Bodies for Medical devices are not connected to this platform because most of the European Union Member Countries have not recognized these devices as measuring instruments in the legal metrology area, although these instruments may have an impact on the end user if the measuring is inadequate. Considering that one of the important parts of legal metrology is health protection, it leads to the fact that individual countries recognized these measuring instruments as instruments for which metrological control is mandatory.

Notified Bodies Association should harmonize the approach to conformity assessment for measuring instruments in European Union, and it should increase the level of trust in measuring instruments, which are placed on the market. Also, communication between NoBoMet and European Organization for Legal Metrology WELMEC should enable better set up requirements, which instruments in legal metrology should fulfill. It can be concluded, that the European Union pays attention for protecting its interests, by implementation requirements from legal metrology.

Medical devices from the aspects of legal metrology are representing the challenge for legal metrology, i.e. putting them into frames of legal metrology in order to protect the health of the end user by providing accurate and undoubted results.

7 Role of Medical Devices in Framework of Legal Metrology

Although medical devices with measuring functions are not covered by the legal regulations from the aspect of initial and subsequent verification in all countries, they have a crucial role in health protection of users by conducting accurate measurements. Considering their widespread use, these devices should be regulated in a similar way as measuring instruments mentioned in NAWI and MID Directives (type approval, initial verification, subsequent verification, metrological control, maximum permissible errors, etc.). Some Members and Associate Members of WELMEC have been recognizing medical devices with measuring function as instruments that need initial and subsequent verification to ensure accurate results or health protection of consumers.

Medical devices in the European Union are regulated by specific directives, with the goal of setting up the essential requirements for these instruments from the aspects of health, safety, and use. Also, the purpose of these directives is reflected through trade barriers removal. Medical devices directives belong to the before mentioned New Approach Directives.

Directives dealing with medical devices are: Directive 90/385/EEC [52], Directive 93/42/EEC [53] and Directive 98/79/EEC [54].

With the appearance of these directives, new responsibilities and technical requirements for a medical device were created before they were put on the market. Also, directives mention and describe the roles of all participants, from manufacturers, distributors/importers, users, society to governments.

Besides regulations mentioned in directives, each European Union Member Country can prescribe additional measures to protect the health of its citizens. Opportunity for performing surveillance of medical devices with measuring function is within the domain of legal metrology, and it is done by initial and subsequent verification of these instruments and by their placement on the market.

As measuring instruments mentioned in the NAWI and MID directive must be marked with a "CE" mark, in the same way, medical devices should be marked before their placing on the European Union market. These marks are set by Notified Bodies from medical devices area.

In order to better understand medical devices, definition for these devices in accordance with Directive 2007/47/EC [55] is given:

Medical device means any instrument, apparatus, appliance, software, material or other article, whether used alone or in combination, together with any accessories, including the software intended by its manufacturer to be used specifically for diagnostic and/or therapeutic purposes and necessary for its proper application, intended by the manufacturer to be used for human beings for the purpose of:

- diagnosis, prevention, monitoring, treatment or alleviation of disease,
- diagnosis, monitoring, treatment, alleviation of or compensation for an injury or handicap,

- research, replacement or modification of the anatomy or of a physiological process,
- and control of conception,

and which does not achieve its principal intended action in or on the human body by pharmacological, immunological or metabolic means, but which may be assisted in its function by such means.

From this medical devices definition, it can be concluded that not all devices have measuring function necessary for those instruments to be considered relevant for health protection from the aspect of accurate and undoubted measurement. Therefore, it is necessary to identify those instruments and introduce them to the legal metrology framework. Some of devices with measuring function could be Defibrillator, Infusomats and Perfusors, Patient monitors, Neonatal and paediatric incubators, Respirators, Anaesthesia machines, Therapeutic ultrasound devices, Dialysis machines, ECG—Electrocardiographs [56], and etc.

From the aspect of legal metrology for medical devices with the measuring function, it is necessary to establish adequate chain traceability for measurement results as it is in the case for measuring instruments covered by NAWI and MID Directive.

Traceability property of a measurement result whereby the result can be related to a reference through a documented unbroken chain of calibrations, each contributing to the measurement uncertainty [57].

Traceability of results is ensured through calibration procedures.

Calibration an operation that, under specified conditions, in a first step, establishes a relation between the quantity values with measurement uncertainties provided by measurement standards and corresponding indications with associated measurement uncertainties and, in a second step, uses this information to establish a relation for obtaining a measurement result from an indication [57].

Measurement standard realization of the definition of a given quantity, with stated quantity value and associated measurement uncertainty, used as a reference [57].

In the context of metrology, "trust" means the traceability of the measurement results to the SI unit with some degree of confidence. In the health sector, because of the absolute risk of human life, it is necessary to measure these quantities as accurately as possible.

The purpose of the measurements in medicine is very important because it is reflected on the prevention of treatment of disease and monitoring patient. Above mentioned measurements are performed on a daily basis so they must be accurate, and the results must be comparable to various locations in time. Only in this way it is possible to optimize care about patient and to put legal metrology in the function of citizen's protection [58].

As an example, if the blood pressure was measured inaccurately, it can have a significant impact on the health of the patient, and the conclusion could be that medical devices with this measuring function should be covered by the legal framework.

Currently, within EURAMET [59] there is a proposed project named **"Traceability in medical measuring"** in which most National Metrology Institutes will participate with the aim to develop standards, which will be traceable to the SI system, this will ensure adequate traceability of medical devices with measuring function [60].

Medical devices can be classified as following [58]:

- Group I—requires conformity assessment,
- Group II a—requires conformity assessment and certification of quality system,
- Group II b—requires conformity assessment and independent testing and certification of the quality system,
- Group III—requires independent assessment of the design, conformity assessment, and independent testing and certification of the quality system.

These modules can be compared to modules mentioned in Measuring Instruments Directive (MID). MID Directive provides a modern approach, which gives much more space for manufacturers in conformity assessment procedures, aligning European Union legislation with international standards. However, by some medical devices with measuring function, adequate metrological traceability of the measuring results has not been established yet, making their application difficult in the field of legal metrology [58].

Some recommendations for medical devices with measuring functions are available within OIML, namely [58, 61]:

- R 16—Sphygmomanometers (E:2002),
- R 26—Medical syringes (E:1978),
- R 40—Standard graduated pipettes for verification officers (E:1981),
- R 89—Electroencephalographs-Metrological characteristics-Methods and equipment for verification (E:1990),
- R 90—Electrocardiographs-Metrological characteristics-Methods and equipment for verification (E:1990),
- R 103—Measuring instrumentation for human response to vibration (E:1992),
- R 104—Pure-tone audiometers (E:1993),
- R 114—Clinical electrical thermometers for continuous measurement (E:1995),
- R 115—Clinical electrical thermometers with maximum device (E:1995), and
- R 122—Equipment for speech audiometry (E:1996).

In order to adequately cover medical devices with measuring function within legal metrology, it is necessary to develop recommendations on European or international level, as it is the case with measuring instruments covered by NAWI and MID Directives. This will require the establishment of working groups within WELMEC and Technical Committee within OIML. New groups will have the aim to create guidelines that will harmonize conformity assessment procedure, initial verification, subsequent verification, and surveillance of these instruments, as well as maximum permissible errors by performing initial and subsequent verification. Also, each Member aiming to become a part of these working groups (country) needs to improve its legislation in the legal metrology area related to medical devices with a measuring function with the purpose of health protection of its citizens. This implies mandatory verification for these instruments in particular time intervals. In the case that instruments do not satisfy limits of the prescribed error they will be put out of the service.

8 Legal Metrology System—Future

In the previous text it was demonstrated that the legal metrology is a part of daily life and that it has the impact on commercial transactions, safety, and environmental protection as well as on health. Considering globalization and increasing networking on the worldwide level implementation of legal metrology will indeed grow, and demands for legal metrology will also increase. Therefore, each country should recognize legal metrology as the crucial part of its society and future development. Identifying legal metrology in this way would lead to adopting regulations, which would manage this area according to the international regulations. Arranging the legal metrology system will be reflected in the way that a particular country would (authorized body) participate in the international or regional organization of legal metrology more intensively with the purpose of exchanging experiences and improvements of their system.

All the regional metrology organizations will have a very important role in the future of legal metrology, particularly the International Organization of Legal Metrology (OIML). Improvement of legal metrology together with regional organizations will be reflected through knowledge exchange, on the development of methods to establish the uniform approach to measuring, as well as measuring instruments that come out of legal metrology and recognition of test reports.

To accomplish successful harmonization of measurement requirements in the future it will be necessary to establish requirements for technical and measuring features, requirements for tests, procedures for testing and format of test reports. This means that in the future it will also be necessary to establish the accuracy class for measuring requirements, maximum permissible error in the nominal and working conditions. Technical requirements will, on the other hand, refer to the validity of measuring instruments and accurate display. Trends in verification are to use remote control systems of the measurement system [4, 62–67].

Development of legal metrology will be encouraged by development of new technologies (R&D) and new technology development will be encouraged by development of the society. This will lead to creating new products such as new tools and instruments that will serve a purpose to developing and creating a new society. Development of society will, on the other hand, lead to the need for metrology development. This means that the metrology will have benefits from the development of new technologies. The new society and metrology will be the creators of legal metrology. New technologies that will appear are: information, environmental protection and biotechnology. Some of these new technologies comprise: intelligent mobile phones with built-in sensors, robots, computers, DNA chips or micro-chips for analytical systems, etc. [62] With the appearance of new technologies and new instruments it will be necessary to set up correct traceability to primary units of the SI system. There will be an overlap of scientific and legal metrology with an aim of dealing with arising issues.

Anyway, the role of the legal metrology system will remain the same and its role is to ensure accurate and undeniable measurement.

Conclusively, measurements that will appear together with the development of global civilization, as well as humans' need for development will for sure remain the same in the future.

References

1. BIPM. http://www.bipm.org/en/worldwide-metrology/, last access date 24.8.2016
2. BIPM. http://www.bipm.org/en/worldwide-metrology/metre-convention/, last access date 24. 8.2016
3. OIML B1 (E:1968)—OIML Convention
4. Chappell S (2004) Opportunities and future trends in legal metrology control of measuring instruments. Organisation Internationale de Métrologie Légale (OIML). Bulletin 55(1):25–28
5. International vocabulary of terms in legal metrology (VIML) E:2013
6. OIML on-line vocabulary, http://viml.oiml.info/en/index.html, last access date 24.8.2016
7. Birch J (2004) The expanding scope of legal metrology and the changing role of the state in a globalization world. Organisation Internationale de Métrologie Légale (OIML). Bulletin 55 (1):23–24
8. Mason P (2016) Re-thinking Legal Metrology Organisation Internationale de Métrologie Légale (OIML). Bulletin 1:14–17
9. Organisation Internationale de Métrologie Légale/ International Organization of Legal Metrology
10. OIML B15 (E:2011)—OIML Strategy
11. OIML, https://www.oiml.org/en/about/legal-metrology, last access date 20.8.2016
12. OIML, https://www.oiml.org/en/about/benefits?set_language=en, last access date 20.8.2016
13. Awosola M (2010) Working across cultures in OIML TCs and SCs. Organisation Internationale de Métrologie Légale (OIML). Bulletin 51(3):17–19
14. OIML. https://www.oiml.org/en/structure/members, last access date 20.8.2016
15. OIML. https://www.oiml.org/en/structure/ciml, last access date 20.8.2016
16. OIML. https://www.oiml.org/en/structure/conference?set_language=en, last access date 20.8. 2016
17. Le Comité international de métrologie légale (CIML)/The International Committee of Legal Metrology
18. Le Bureau international de métrologie légale (BIML)/The International Bureau of Legal Metrology
19. OIML. https://www.oiml.org/en/technical-work/tc-sc/tclist_view, last access date 20.8.2016
20. OIML. https://www.oiml.org/en/structure/biml, last access date 20.8.2016
21. WTO Technical Barriers to Trade (TBT) Committee
22. Kochsiek M, Andreas O (2001) Towards a global measurement system: contributions of international organizations. Organisation Internationale de Métrologie Légale (OIML). Bulletin 42(2):14–19
23. Basic Certificate System for Measuring Instruments

24. OIML B3 (E:2011)—OIML Basic Certificate System for OIML Type Evaluation of Measuring Instruments
25. OIML Systems Basic and MAA Certificates (2016) OIML Bulletin 67(2):39
26. OIML B10 (E:2011)—Framework for a Mutual Acceptance Arrangement on OIML Type Evaluations (Integrating the changes in the 2012 Amendment)
27. ISO/IEC 17065—Conformity assessment—Requirements for bodies certifying products, processes and services
28. ISO/IEC 17025—General requirements for the competence of testing and calibration laboratories
29. OIML. https://www.oiml.org/en/certificates/maa-certificates, last access date 20.8.2016
30. WELMEC. http://www.welmec.org/welmec/welmec-tour/welmec-information/, last access date 21.08.2016
31. The European Free Trade Association (EFTA) is an intergovernmental organization set up for the promotion of free trade and economic integration to the benefit of its four Member States: Iceland, Liechtenstein, Norway, Switzerland
32. WELMEC. http://www.welmec.org/welmec/welmec-tour/welmec-organization/, last access date 22.8.2016
33. Directive 2014/31/EU of the European Parliament and of the Council of 26 February 2014 on the harmonization of the laws of the Member States relating to the making available on the market of non-automatic weighing instruments (recast). Applicable on 20 April 2016 OJ L 96, 29 March 2014
34. Directive 2014/32/EU of the European Parliament and of the Council of 26 February 2014 on the harmonization of the laws of the Member States relating to the making available on the market of measuring instruments (recast). Applicable from 20 April 2016 OJ L 96, 29 March 2014
35. NIST. http://www.nist.gov/pml/wmd/ilmg/org-primer.cfm, last access date 21.08.2016
36. SIM. http://www.sim-metrologia.org.br/, last access date 21.08.2016
37. APLMF. http://www.aplmf.org/, last access date 21.08.2016
38. COOMET. http://www.coomet.net/, last access date 21.08.2016
39. Directive 93/68/EEC. http://eur-lex.europa.eu/legal-content/en/ALL/?uri=CELEX:31993L0068, last access date 23.8.2016
40. WELMEC. http://www.welmec.org/welmec/country-info/bosnia-and-herzegovina/, last access date 23.8.2016
41. Oosterman C, Dixon P (2012) NoBoMet: a European platform for notified bodies working in legal metrology. Organisation Internationale de Métrologie Légale (OIML). Bulletin 53 (1):11–13
42. Directive 765/2008 EC amends Directive 768/2008/EC
43. Nando (New Approach Notified and Designated Organizations) Information System,
44. http://ec.europa.eu/growth/tools-databases/nando/, last access date 23.8.2016
45. OIML V1, International vocabulary of terms in legal metrology (VIML) E:2013 Directive 768/2008/EC. http://eur-lex.europa.eu/legal-content/EN/TXT/?qid=1471938874347&uri=CELEX:32008D0768, last access date 23.8.2016
46. Netherlands Measurement Institute (NMI)
47. Czech Metrology Institute (CMI)
48. Laboratoire national de métrologie et d'essais (LNE)
49. National Measurement and Regulation Office (NMO)
50. Swiss Federal Institute of Metrology (METAS)
51. NoBoMet. https://sites.google.com/site/portalnobomet/home, last access date 23.8.2016
52. European Commission. https://ec.europa.eu/growth/single-market/european-standards/harmonised-standards/implantable-medical-devices_en, last access date 24.8.2016
53. European Commission. https://ec.europa.eu/growth/single-market/european-standards/harmonised-standards/medical-devices_en, last access date 24.8.2016
54. European Commission. https://ec.europa.eu/growth/single-market/european-standards/harmonised-standards/iv-diagnostic-medical-devices_en, last access date 24.8.2016

55. European Commission. http://ec.europa.eu/consumers/sectors/medical-devices/files/revision_docs/2007–47-en_en.pdf, last access date 24.8.2016
56. WELMEC. http://www.welmec.org/welmec/country-info/bosnia-and-herzegovina/, last access date 24.8.2016
57. International Vocabulary of Metrology—Basic and General Concepts and Associated Terms (VIM 3rd edition), JCGM 200:2012
58. do Céu Ferreira M (2011) The role of metrology in the field of medical devices. Organisation Internationale de Métrologie Légale (OIML). Bulletin 2(2):135–140
59. European Association of National Metrology Institutes (EURAMET)
60. EURAMET. http://msu.euramet.org/current_calls/research_2016/SRTs/SRT-r03.pdf, last access date 24.8.2016
61. OIML. https://www.oiml.org/en/publications/recommendations/publication_view?p_type=1&p_status=1, last access date 24.8.2016
62. Tanaka M (2004) Measuring instrument technology and customers and contractors of legal metrology in the mid 21st century. Organisation Internationale de Métrologie Légale (OIML). Bulletin 55(1):29–31
63. Badnjevic A, Gurbeta L, Boskovic D, Dzemic Z (2015) Measurement in medicine—past, present, future. Folia Medica Facultatis Medicinae Universitatis Saraeviensis Journal 50 (1):43–46
64. Badnjevic A, Gurbeta L, Boskovic D, Dzemic Z (2015) Medical devices in legal metrology, IEEE 4th Mediterranean Conference on Embedded Computing (MECO), pp 365–367, 14–18 June, Budva, Montenegro
65. Gurbeta L, Badnjevic A, Pinjo N, Ljumic F (2015) Software package for tracking status of inspection dates and reports of medical devices in Healthcare Institutions of Bosnia and Herzegovina, XXV International Conference on Information, Communication and Automation Technologies (IEEE ICAT), pp 1–5, 29–31 October 2015, Sarajevo, Bosnia and Herzegovina
66. Gurbeta L, Badnjevic A, Sejdinovic D, Alic B, Abd El-Ilah L, Zunic E (2016) Software solution for tracking inspection processes of medical devices from legal metrology system, XIV Mediterranean Conference on Medical and Biological Engineering and Computing (MEDICON), 31. March—02. April 2016. Paphos, Cyprus
67. Gurbeta L, Badnjevic A (2016) Inspection process of medical devices in healthcare institutions: software solution. Health Technol. doi:10.1007/s12553-016-0154-2

Part I
Inspection and Testing of Diagnostic Devices

Inspection and Testing of Electrocardiographs (ECG) Devices

Ratko Magjarević and Almir Badnjević

Abstract Electrocardiographs are nowadays standard part of diagnostic procedure in healthcare systems since they have high significance in diagnosis of large number of diseases and disorders. Due to development of technology, especially electronics, these devices have been revolutionized since the first prototype was invented. Nowadays, these devices are able to perform automated diagnosis and measure multiple parameters at once. Multiple international standards define device life circle, from production to disposal. However, this sophistication of ECG devices raises numerous questions regarding safety and accuracy. This chapter describes basic principles of electrocardiography and ECG devices as well as gives overview of requirements in area of safety and performance inspection of these devices.

1 Historical Aspects of Electrocardiography

Electrocardiography (ECG) is the method of measuring of electrical potentials of the heart in order to discover heart related health problems. The recordings of the potentials of the heart (on paper or on other media) are called the electrocardiogram and the medical device used for the recording are electrocardiographs (the same abbreviation ECG is used for all three terms). The potentials of the heart are recorded against time. The procedure is undertaken from the body surface during standard check-ups, when only a few seconds of the ECG are printed on paper or viewed on a monitor. However, in cases of patients suffering from a heart disease, recordings may be taken for a much longer period, e.g. during emergency or in

R. Magjarević (✉)
Faculty of Electrical Engineering and Computing, University of Zagreb,
Zagreb, Croatia
e-mail: Ratko.Magjarevic@fer.hr

A. Badnjević
Medical Devices Verification Laboratory, Verlab Ltd Sarajevo, Sarajevo,
Bosnia and Herzegovina
e-mail: badnjevic.almir@gmail.com

© Springer Nature Singapore Pte Ltd. 2018
A. Badnjević et al. (eds.), *Inspection of Medical Devices*, Series in Biomedical
Engineering, https://doi.org/10.1007/978-981-10-6650-4_4

intensive care. In case the abnormality in the heart potential appears rarely, over a period of time longer than a few hours, the ECG is recorded by an ECG holter, typically for 24 h. Digitized signals are stored in a memory, for further computerised analysis. For the recording, electrodes are placed in standardised positions on a patient's body. An ECG is an important part of every preventive medical check-up and it is mandatory for assessment of a patients who are suspected to have a heart related problem. The ECG is considered as an extremely safe procedure and without any risk involved. Adverse effects reported in rare cases, deal with skin irritation from the electrode adhesive. Though ECG is an important procedure in evaluation of cardiac patients, additional examinations are undertaken to get the full picture of symptoms and the disease. In some cases, normal recordings are obtained in patients with heart disease, or some recorded parts of ECG may be recognised as pathological despite the heart is in normal condition. The ECG provides information about the heart's electrical activity and has a great value in finding the causes of symptoms like chest pain or pressure. It is used for interpretation of the severity of a heart attack, inflammation of the pericardium, angina or other symptoms of heart disease like shortness of breath, dizziness or even fainting, and of arrhythmias. By ECG interpretation, physicians may conclude on some physical dimensions of the parts of the heart, e.g. the thickness of the heart chambers walls. ECG reflects the efficiency of medication and enables finding of their side effect. Surface ECG is also used in regular control of implanted devices for heart management, like pacemakers and cardioverters—defibrillators (though modern implantable devices enable telemetric measurement of intra-cardiac ECG). It is also used in assessment of the health of the heart in presence of other diseases or conditions, e.g. high blood pressure, high cholesterol, diabetes, a family history of early heart disease or history of smoking [1, 2].

The first observations of effects of electricity on animal tissues happened in late 17th and early 18th century. The Italian physicist and physician Luigi Galvani noted that touching dissected frog's leg with a metal scalpel causes their twitches and explained it with "animal electricity". Later he showed muscle contraction when contacting them to an electrical generator. A very sensitive device for measurement of small voltages and currents is named a "galvanometer" in his honour. Galvani's research was continued by another Italian scientist and inventor, Alessandro Volta, who showed twitching of frog leg muscles due to electrical current generated by plates of two dissimilar metals set against the muscles.

Willem Einthoven (1860–1927), a Dutch physician and physiologist, introduced to medicine the first practical electrocardiograph based on spring electrometer and introduced the term electrocardiogram for changes in the observed potentials [3], Fig. 1. One may today discuss how really practical that device was, since it had a mass of approx. 270 kg and for recording, the examinee was asked to immerse his hands and a leg into containers with salt water which were serving as electrodes. That formation is still used in routine ECG recording named the I, II and III standard limb leads, while the imaginary triangle these measurement points build is named Einthoven's triangle. Einthoven introduced the letter nomenclature for those five deflections which can be recognised in the ECG: P, Q, R, S and T (and later

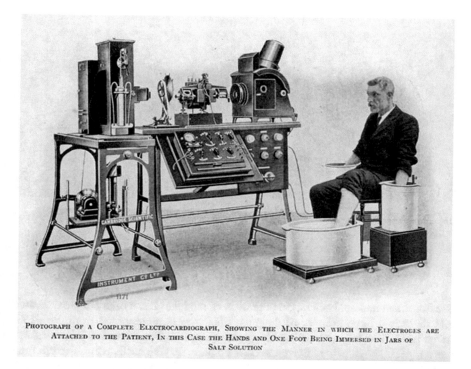

PHOTOGRAPH OF A COMPLETE ELECTROCARDIOGRAPH, SHOWING THE MANNER IN WHICH THE ELECTRODES ARE ATTACHED TO THE PATIENT, IN THIS CASE THE HANDS AND ONE FOOT BEING IMMERSED IN JARS OF SALT SOLUTION

Fig. 1 An early commercial ECG machine, built in 1911 by the *Cambridge Scientific Instrument Company* [4]

also the U wave). In his publication from 1906, Einthoven described normal and abnormal electrocardiograms recorded by the string galvanometer. In 1924, he was awarded the Nobel Prize in Physiology or Medicine for his discovery of the mechanism of the electrocardiogram [3].

In order to make ECG recording practical, a lot of technological improvements were necessary. Amplifiers with vacuum tubes were introduced in 1928 by Ernstine and Levine [5], and later cathode ray tube for displaying the potentials on the screen. Electrocardiographs designed in analogue technology were equipped with chart recorders, where the first models were writing with ink on grid paper and later using hot wire on temperature sensitive paper. Containers with salt water were replaced with dry electrodes only in 1930s by silver plate electrodes and by suction electrodes, normally used for recording the precordial leads [6]. Early ECGs used vacuum tubes and were therefore heavy, unreliable and they had large power consumption. Invention of silicon transistor in 1947 enabled production of smaller and more practical ECG devices and facilitated diagnostic use of the ECG, Fig. 2.

(a)   **(b)**

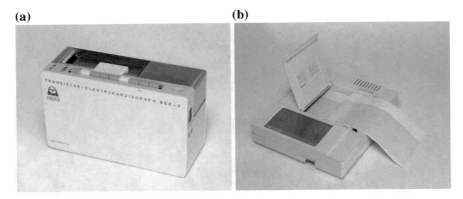

Fig. 2 **a** Single channel ECG—Transistor—Elektrokardiograph BEK-3 from 1970 and **b** Multichannel ECG Siemens Burdick E 550 approx. 1995

2 Medical Aspects

Measurement and analysis of the potentials of the heart has a high diagnostic value and there is a large number of diagnoses that can be determined from the ECG, much more than from any other bioelectric potential. The normal ECG waveform has a regular shape, where five characteristic parts may be easily recognised in time in nearly every standard lead records: P-wave, QRS complex and T-wave. The ECG signal differs from person to person, but it has quasi stationary behaviour. The spectrum of the ECG is characterised mainly by the shape of the five characteristic parts, their time relations within the cardiac cycle and by the variability of the hearth rhythm. The analysis of the ECG enables determination of most irregularities and arrhythmias of the heart muscle.

The heart is a hollow muscle organ consisting of four chambers. The heart is filled with blood and enables the circulation of the blood in the cardiovascular system by regular rhythmic contractions. The heart receives the blood from the veins and pumps it out into the arteries, Fig. 3. The right and the left part of the heart consist each of two chambers, the atrium and the ventricle, and they are separated by a muscle called septum. The left ventricle is the strongest muscle of the heart and it pumps the blood to the largest vessel—the aorta. The directed blood flow in the cardiovascular system is regulated by the rhythmic contractions of the chambers, firstly by synchronous contraction of the atria and then by synchronous contraction of the ventricles. Backflow (reflux) of the blood from ventricles to atria, and from arteries to ventricles, is disabled by four heart valves. The valves are positioned as follows: in the right heart, the tricuspid valve is between the atrium from the right ventricle, and the pulmonary valve between the right ventricle and the pulmonary arteries, whereas in the left heart, the mitral valve is between the left atrium and ventricle and the tricuspid valve between the left ventricle and the aorta.

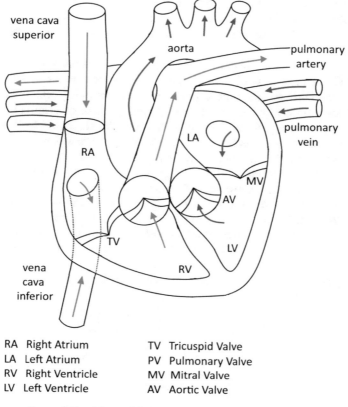

vena cava superior

aorta

pulmonary artery

pulmonary vein

LA

RA

MV

AV

TV

LV

vena cava inferior

RV

RA	Right Atrium	TV	Tricuspid Valve
LA	Left Atrium	PV	Pulmonary Valve
RV	Right Ventricle	MV	Mitral Valve
LV	Left Ventricle	AV	Aortic Valve

Fig. 3 Cross-section and blood flow of the heart

The pumping phase of the heart cycle is designated as the systole, and the resting phase as diastole. The contraction of the heart muscles is ruled by precise electrical activity of a specialized conduction system of the heart. The normal cardiac cycle is initiated by activity of special cells positioned on the top of right atrium, at the sinoatrial node (SA) (Fig. 4). These cells are the natural pacemaker of the heart. From the SA node, the depolarization spreads over the whole atria, causing the contraction of atrial tissue. Since the atria and ventricles are separated by a fibrous ring with low conductivity, the electrical depolarization from atria enters ventricles through the atrioventricular (AV) node which generates a short delay in spreading the depolarization. From the AV node, the depolarization spreads by the bundle of His through the left and the right bundle to the left and the right ventricle. In normal rhythm, the contractions of the left and the right heart happen simultaneously, enabling the heart to efficiently pump out the blood into the arteries. The heart is positioned in the thorax between the lungs, protected by the pericardium, the breastbone and the ribs, and covered by a thin layer of muscles that enable breathing and with the skin. The electrical properties of these tissue are different between themselves but as a whole they act as a low pass volume filter so that the potentials recorded at different positions on the heart significantly differ from the surface ECG (Fig. 4).

The electrical activity of the myocardium (cardiac muscle) is reflected in the surface ECG, Fig. 5. In normal ECG, the P-wave represents the depolarisation of the atria. The QRS complex the depolarisation of ventricles, where the repolarisation of the atria is covered by the much stronger signal of ventricular depolarisation. The T-wave represents the repolarisation of the ventricles. In interpretation of the ECG, the S-T segment has an important role since its waveform, in particular the deviation from the baseline, may designate a serious damage of the heart muscle, e.g. the elevation of the ST-segment may designate acute myocardial infraction. The interval between two R peaks, R-R interval, is often used for calculation of the heart rate since it is its reciprocal value. The normal heart rhythm (also called normal sinus rhythm) lies between 50 and 100 beats per minute (in adult healthy person). Normal heart rate in resting varies a little, and the variation is larger since it adopts to the activity of the person, Fig. 6. Deviations from normal heart rate are called arrhythmias: slow heart rate is called bradycardia and fast heart rate tachycardia. Sustainable long-term tachycardia may lead to ventricular

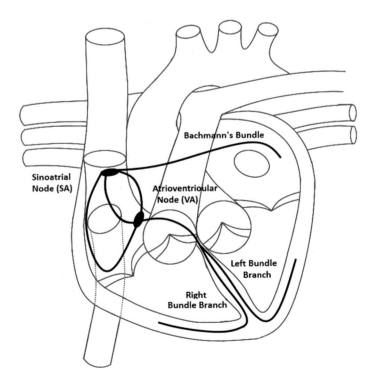

Fig. 4 Conduction system of the heart

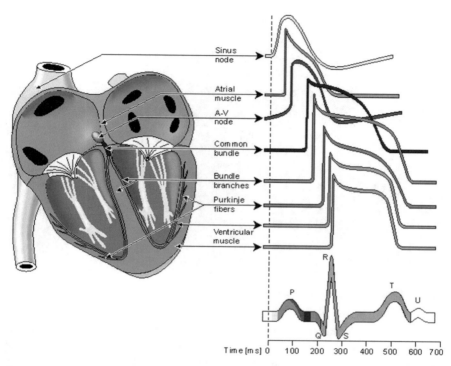

Fig. 5 Action potentials at different positions within the heart [6]

fibrillation which, in case it is not stopped timely, leads to the death. Since in ventricular fibrillation there is no regular, and synchronized contraction of muscle fibers, the blood flow stops and the oxygen supply of the brain is interrupted. Other frequent deviation from regular heart rhythm includes appearance of ectopic beats, generated by cells in myocardium with lower excitation threshold, which fire asynchronously as referred to the normal heart rhythm and do not contribute to the heart output since they are premature. Interruption in conducting of the pulses from atria to ventricles is called atrioventricular block and leads to absence of ventricular contraction. It can be recognised as missing R peak after a P-wave.

2.1 ECG Electrode Placement

Electrodes are the necessary interface for connecting human body to medical electronic instrumentation. In case of ECG recording, the electrodes can be attached to the surface of the body or implanted. In this chapter, authors consider only recording of surface ECG, so only surface electrodes will be described.

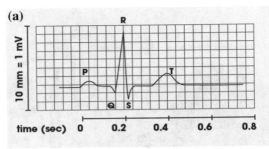

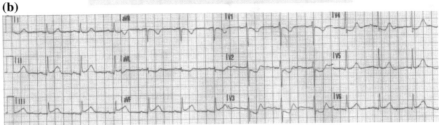

Fig. 6 Characteristic ECG wave shape **a** symbolic and **b** measured

ECG electrodes are transducers which enable exchange of charge carriers in the system consisting of the human body and the medical device. In the body, charge carriers are ions of both polarities, and in the electronic equipment, the carriers are electrons. Though electrodes seem to be simple in their design, they may cause a lot of interference and noise at the interface. More on technical characteristics of electrodes is presented in the technical description of the ECG equipment later in this chapter.

From the point of interpretation of the ECG, physicians evaluate the recorded curves by comparing them to the recordings they have previously seen and used in their training, and from the experience they got practicing medicine. The American College of Cardiology Foundation stated that it takes 3,500 supervised ECG reads to become an expert [7]. Since the ECG records vary between individuals, a lot of skill is necessary to observe the common in ECG features. Placement of recording electrodes always to the same position on patient's body establishes a reference that helps in observing those common characteristics that lead to accurate diagnostics. For practical reasons, the electrodes for standard ECG recording are positioned on human extremities being easily accessible and the housing of electrodes is colour coded so that they enable rapid connecting to a medical device for e.g. recording in emergency cases, Fig. 7.

Standard ECG recording presents bioelectric potentials of the heart recorded in 12 traces and by 10 electrodes positioned on the body [1]. The potentials measured in-between the electrodes are called leads. The leads can be explained as projections of the heart vector to different planes on the body, as shown in Fig. 8. All leads together represent the heart vector projection in practically all directions, which

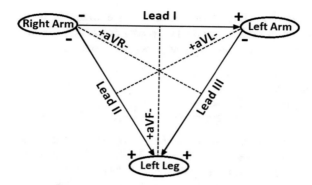

Fig. 7 Standard electrode placement

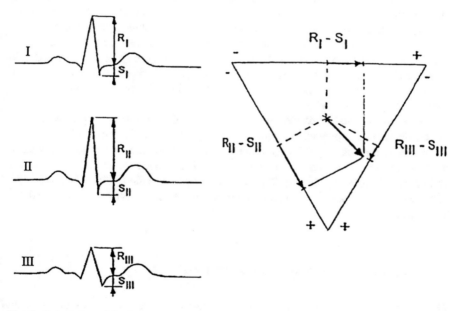

Fig. 8 Calculation of the heart vector

means that they the information on electrical activity of the heart is acquired with high spatial resolution. Those 12 leads are organized as follows:

- Three bipolar limb leads,
- Three augmented limb leads,
- Six precordial or chest leads.

Four electrodes are placed on the all four extremities of the human body (therefore also called limb electrodes) on predefined places slightly proximal to the hand and the ankle. These positions are designated by:

- LA standing for left arm,
- RA standing for right arm,
- LL standing for left leg and
- Ground for the reference electrode, positioned on right leg.

The limb electrodes are used for bipolar or differential measurement of the potentials between them and they form the I, II and III standard leads of the ECG with the orientation of the polarity as shown in Fig. 7. These leads are also called Einthoven leads since they are based on the first, historical ECG measurement performed by Einthoven.

The same limb electrodes are used for augmented leads which are unipolar measurements where the potential present at a particular electrode is measured against the average value of the potentials from the other two electrodes. The augmented leads are also called Goldberger's leads in honour of Emanuel Goldberger who introduced the leads to electrocardiography in 1942 in order to increase the recorded voltage aginst Wilson's central terminal. The projections of the heart vector of augmented leads are shown in Fig. 7. The augmented leads are marked aV_R, aV_L and aV_F.

Precordial or chest leads measure the potential of from the electrodes positioned on the rib cage of a patient in a predefined way against the potential of a reference electrode. The potential of the reference electrode is defined by connecting each of the three limb electrodes through a resistor of 5 kΩ into the reference node which is considered to have 0 potential. The reference node is called Wilson's central terminal or electrode and the precordial leads got the name Wilson's leads and are labelled as V_1–V_6.

2.2 Bipolar Leads and Einthoven's Law

The potentials measured between the limb electrodes are projections of the heart vector to the Einthoven equilateral triangle. Einthoven's Law states that the electrical potential of any limb equals the sum of the other two. The electrodes and the polarity when the ECG is measured by the standard limb leads is the following (Fig. 8):

- Lead I—The negative terminal of the ECG amplifier is connected to the right arm, and the positive terminal is connected to the left arm.
- Lead II—The negative terminal of the ECG amplifier is connected to the right arm, and the positive terminal is connected to the left leg.
- Lead III—The negative terminal of the ECG amplifier is connected to the left arm, and the positive terminal is connected to the left leg.

Einthoven's Law states that the electrical potential of any limb equals the sum of the other two (+ and − signs of leads must be observed).

2.3   Cardiac Vector

Vectors are used to describe depolarization and repolarization events of the heart muscle. Cardiac vector shows the direction of charge spread in the heart muscle as they happen in time and the magnitude of the electrical activity (Fig. 8). Each lead of the ECG represents a projection of the cardiac vector into different plane. The standard three leads and the augmented leads show each the projection of cardiac vector into a plane which is 60° rotated from the previous lead (Fig. 7). In case there is no electrical activity of the heart, the projection is zero and the ECG shows the baseline, a horizontal line.

Standard record of an ECG is on paper or on the screen of a monitor. In order to make it adopted for measurement in time (duration of ECG waves, segments etc.) the printing is calibrated so that 1 mm equals 40 ms at standard paper speed 25 mm/s. In case better time resolution is needed for diagnostics, paper speed is increased to 50 mm/s. The electrical axis of the heart is a sum of all vectors activated in a particular chamber of the heart. In such a way, each part of the ECG has its own respective vector.

2.4   Recording of an ECG

Since the ECG is one of the most informative diagnostic examination of the heart and cardiovascular system, it became a routine procedure in screening patients.

The routine procedure is taken with the patient laying on a bed. The patient's skin on the positions provided for electrode placement on the chest and the extremities are cleaned and the electrodes attached. In standard procedure, suction electrodes are attached to the chest and clip electrodes to the arms and legs. The electrodes are connected to the cables of the electrocardiograph and after the patient calms down, the recording begins. The patients have o remain still during the recording. They may be asked to hold breath for a few seconds. Any movements during the recording may introduce artefacts which degrade the potentials of the heart. In some patients, the diagnostic procedure is specified as a stress test, and then it is recorded under controlled exercising. The recorded electrocardiograms are reviewed by physicians.

Patients taking any kind of medication should inform the physician. They should not be physically active before the test. The procedure itself is comfortable and patients feel well during and after the test.

The diagnostic value of the ECG primarily in screening of cardiac arrhythmias and abnormalities of the conduction system of the heart, as well as in detecting myocardial ischemia. The ECG is used for monitoring of drug intake and of the performance of implanted devices like pacemaker. The ECG is also used in interpretation of hypertension, cardiomyopathy, valvular disease, metabolic diseases and many others.

3 ECG Devices—Technical Description

An electrocardiograph (ECG) is a measuring device and it is designed as an open measurement channel. The main parts of the ECG are: a set of electrodes, a lead selector, an amplifier, filters, a printer and/or a display unit. The ECG may be designed as a 3 channel, 6 channel or a 12 channel device, though for a number of years 12 channel device dominate since they match to 12-lead standard in ECG interpretation. The device is able to record bipolar, augmented and precordial leads. Many multichannel electrocardiographs acquire and analyse the ECG signals since they have embedded microprocessors with ECG signal processing software. The signals recorded at the surface have a range of magnitude of 1 mV and the spectrum between 0.05 and 150 Hz [8]. The block diagram of an analogue front end of an electrocardiograph is presented in Fig. 9. In Fig. 10, a block diagram of an integrated circuit with ECG capabilities is presented. Immediately after amplification, the signals are digitized and further processed digitally. Front-end sampling may be performed at rates from 1000 to 2000 samples per second. Active right leg drive (RDL) is integrated to the circuit as well. Detailed descriptions of the functionalities of the ECG integrated circuits may be found at the Internet sites of leading integrated circuits producers, Fig. 10.

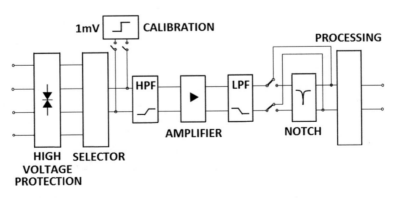

Fig. 9 Block diagram of an electrocardiograph with analog front end

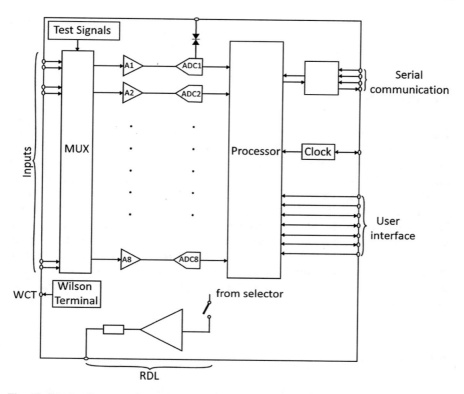

Fig. 10 Block diagram of an integrated circuit comprising many functionalities of an electrocardiograph

3.1 Electrodes

Electrodes for recording biopotentials are usually produces of an inert metal. In ECG, most commonly silver-silver chloride (Ag-AgCl) or stainless steel are used due to their biocompatibility and chemical stability. Each metal electrode that is in contact with electrolyte (which is always present in biological tissue), produces a double layer where positive ions are attracted to the electrode surface and negative ions forms adjacent to it. The potential drop over the double layer is called half cell potential, which changes in time with changes of ion concentration and temperature. The characteristic of Ag-AgCl electrodes is a constant half cell potential of approximately 0.8 mV. Clip and suction electrodes are usually reusable while selfadhesive ECG electrodes are expendable. They carry a conductive gel soaked sponge under the snap which is applied to the skin once the electrodes are attached into the position for recording. The conductive gel lowers the resistance of the skin under the electrode which contributes to the quality of the recording because it is lowering the sensitivity of the input stage to interference. Conductive gel may be applied to the skin after cleaning out of the same reason [8].

3.2 Lead Selector

Electrodes are connected to the input stage of the ECG by shielded cables with a standardised electrode snap on the electrode end. In the input stage of the device, overvoltage protection circuit in built-in immediately at the front end. The selector has the function to connect the signals from the electrodes to the input of the particular amplifier, directly or through a resistor network so that the selected leads are amplified, processed and displayed according to the number of channels of the ECG. The functionality of the selector is realised by a multiplexer. An important part of the input stage is a calibrator which has to be connected to the input stage in order to ensure that all parts of the measurement channel are calibrated. The calibration of ECG devices is provided by connecting a 1 mV step function to the input of each measurement channel. For devices with printers and paper with millimetre grid, the deflection for 1 mV input voltage should be calibrated to 10 mm [8].

3.3 Amplifier

Amplifiers for bioelectric signal must have high sensitivity due to low amplitudes of the original signals. High amplification is achieved by multiple stages of amplification in the measurement channel, though the realisation of these stages is mainly within integrated circuits. For amplification of ECG signals, in the input stage, instrumentation amplifiers are used most commonly. Instrumentation amplifier have very high amplification (100–120 dB), very high input impedance (10 MΩ) and a symmetrical structure of the input stage, which all enables realization of a high common mode rejection ratio (CMRR) of the amplifier. CMRR is a measure of suppression of common mode voltage, compared to amplification of the useful bioelectric signals and should be above 100 dB for biopotential amplifiers. The bioelectric signals are prone to electromagnetic interference, especially from the electrical power lines (230 V/50 Hz in Europe) which are superimposed to the useful ECG signal [2].

The signal at the input of the ECG measurement chain consists of four fractions:

1. The measured bioelectric potential (ECG), considered to be the useful signal, with an input range from 50 μV to 1.5 mV.
2. Polarisation voltage which is the difference between the half-cell potentials of two electrodes, ergo, a DC voltage up to 300 mV. The appearance of the polarisation voltage is avoidable whenever the skin is in contact with metal electrodes.

3. Interference from the AC mains frequency (50 Hz or 60 Hz) from line voltages appears as a common mode signal with amplitudes up to 100 mV. Human body, electrodes and connection cables act as an antenna also for signals with higher frequencies but those signals are usually much easier to filter out due to limited frequency range of ECG amplifiers.
4. Interference high voltages that appear at the input of an ECG device are mainly caused by defibrillator shock voltages or by RF surgery equipment. The defibrillator shock can be treated as a single event and the energy of the shock is always known—up to 400 J. The voltages generated by the defibrillator may reach a few thousand volts, but have limited duration. However, the shocks may be repeated several times. Electrosurgical RF devices produce voltages up to a few hundred volts at a frequency between 500 kHz and 5 MHz, but the duration of the application of the voltage through the body is much longer as compared to the defibrillator shock [9]. The protection circuits that are built into the input stage of the ECG protect the internal circuits and the patient from those potentially dangerous voltages.

From the above analysis of the complex signal that can appear at the input of the ECG measurement chain, it is easy to conclude that the bioelectric signal is the weakest and therefore the processing strategy for the input stage has to be well deliberated.

In order to protect the patient from the mains voltage and the other potentially dangerous voltages coming from the mains, and also to protect the device from overvoltage potentially appearing at the patient side e.g. due to defibrillation, the ECG amplifier is in many designs realised as an isolation amplifier. The isolation circuit separates the patient side and the device side with an isolation barrier which can withstand an electrical shock up to 7.5 kV.

3.4 Filters

The frequency range of diagnostic ECG signal spans from 0.05 to 150 Hz thus the instrumentation in the signal processing channel has to band-pass those parts of the ECG spectrum. The electronic filtering circuits for low pass and high pass filter are designed separately. The high pass filter is applied to remove the polarization voltage (DC component) which may be two orders of magnitude larger than the ECG itself and could drive the input amplifier into saturation, and the very slow components of the signal which correspond to wondering of the baseline. The low pass filter removes artefacts from muscle activity and any high frequency interference. Some parts of the myoelectric spectrum overlap with the spectrum of ECG so that remains of the EMG signal can be observed in the records. Also some of the movement artefacts cannot be efficiently removed from the ECG since they spectrums overlap.

The interference from the mains voltage may be eliminated by a notch filter with the central frequency at 50 Hz or 60 Hz. However, due to narrow frequency range of notch filters, analog notch filters have a pronouncedly non-linear phase characteristic, and may change the wave shape of the ECG trace significantly, which may lead to wrong interpretation in reading of the ECG record [10]. Filtering of power noise from ECG is performed mainly by high order linear phase digital filters where the design keeps flatness of the band pass characteristic.

3.5 Display Device

Contemporary electrocardiographs have low power displays and in most cases an embedded printer or wireless connection to a network printer. Digital records may be stored in an appropriate database as a part of the electronic health record of the patient. Electrocardiographs equipped with only paper printers are very rare today.

3.6 Final Assembly

The components of the electrocardiograph are assembled and placed into an appropriate metal frame. The finished devices are then put into final housing along with accessories such as spare electrodes, printout paper, and manuals. They are then sent out to distributors and finally to customers.

4 Safety Aspects

Legal and technical requirements for all those who manufacture and design electromedical devices are numerous and may be dependent on the particular requirements in different regions. In Europe, the legal aspects are regulated by the Medical Device Directive (Council Directive 93/42/EEC of 14 June 1993. The Directive intends to harmonise the national laws that relate to medical devices within the European Union and it is a "New Approach" Directive which means that safety aspects relay on Harmonized Safety Standards for the products, including medical electrical equipment. Devices meeting "harmonised standards" are considered to meet the conformity to the Directive which is confirmed by issuing of the CE mark by a EU Notified Body. The Directive was amended by the 2007/47/EC and the revised directive became mandatory in EU on March 21, 2010. In April 2017, a new regulation on medical devices was adopted in the European Parliament—Regulation (EU) 2017/745. The Commission claims that Regulation brings more consistency into legislation covering safety of medical devices but also adaption of significant technological and scientific progress occurring in the sector in recent past. The new regulation on medical devices will be applied after a transitional period of three years, i.e. in 2020.

The "New Approach" takes standards, which are technical specifications defining requirements for products, production processes, services or test-methods as the measure for safety of products. The specifications are voluntary, but they were developed by stakeholders and they are following the same principles: consensus, openness, transparency and non-discrimination. The standards ensure not only the safety but also interoperability, so important in todays' connected health services.

The European Committee for Standardization (CEN) and the European Committee for Electrotechnical Standardization (CENELEC) closely collaborate with the International Organization for Standardization (ISO) and the International Electrotechnical Commission (IEC) in order to promote the benefits of the international standardisation for trade and market harmonization. The organizations cooperate based on the Vienna Agreement signed by CEN and ISO and the Frankfurt Agreement between CENELEC and IEC. Many of standards in safety of medical equipment created by these European and International organizations are mutually adopted which can be recognised by e.g. IEC/EN marking in the specification of the particular standard.

The Technical Committee (TC) 62 of the IEC, Electrical equipment in medical practice, prepares international standards concerning electrical equipment, electrical systems and software used in healthcare and their effects on patients, operators, other persons and the environment. IEC 60601 series of technical standards deals with the safety and effectiveness of medical electrical equipment. First standards in this series were published in 1977. In the series, there is a general standard IEC 60601-1 that applies to all electrical medical equipment, some 10 collateral standards which are applied more selectively depending on the topic and about 60 particular standards which apply to specific medical equipment, e.g. electrocardiographs. When designing and testing the equipment standardised by a particular standard, both, general and particular standards have to be obeyed. Full text of above mentioned Directives and Standards is available on-line [11–20].

The standard IEC 60601-1:2005 *Medical electrical equipment—Part 1: General requirements for basic safety and essential performance*, describes requirements for basic safety and essential performance applicable generally to all medical electrical equipment. A collateral standard, take IEC 60601-1-2:2014 *Medical electrical equipment—Part 1-2: General requirements for basic safety and essential performance—Collateral Standard: Electromagnetic disturbances—Requirements and tests*, as an example, describes as well basic safety and essential performance of medical equipment but in the presence of electromagnetic disturbances as well as electromagnetic disturbances emitted by medical equipment. Particular standards from 60601 series that deal with electrocardiographs are:

- IEC/EN 60601-2-27 Medical electrical equipment—Part 2-27: Particular requirements for the basic safety and essential performance of electrocardiographic monitoring equipment

– IEC/EN 60601-2-47 Medical electrical equipment—Part 2-47: Particular requirements for the basic safety and essential performance of ambulatory electrocardiographic systems
– IEC/EN 60601-2-51 Medical electrical equipment—Part 2-51: Particular requirements for safety, including essential performance, of recording and analysing single channel and multichannel electrocardiographs

5 Verification and Calibration

Even though different international standards and regulative prescribe all steps in device life cycle, nowadays more attention is given to safety and performance inspections or verifications of these devices during normal usage. These procedure consist of visual inspection of the device, safety inspection in accordance with IEC 60601 and performance evaluation also in accordance with IEC 60601 and manufacturer's recommendations. The purpose of these inspections/verifications is to ensure that device performance during usage is still in stated limits. At this point, metrology as science of measurements is introduced into medical device management. These inspections should be periodical, recommendations is once a year, like preventive check-ups that are basic practice of medical device management in most of the healthcare institutions in the world.

For performing verification/inspection of devices, etalons should be used. For safety inspection there are electrical safety analyser available in the market that allow safety inspection according to IEC 60601, but also in accordance to vast number of electrical safety standards. These analyzers are portable and easy to use, and usually they have software support.

For performance inspections, various analysers can be used also. Generally these analyser's comprise of slots for connecting ECG leads, casing, battery, power supply, user interface. They are often supported with software that enables generation of different performance ECG tests.

The accuracy of the ECG depends on the condition being tested. ECG devices must be constructed and made in a proper way so that in normal working conditions there is protection from electric shock, too high temperature, dust and water into the housing of instrumentation. Reference conditions for ECG:

– Voltage 220–240 V AC, 50 Hz
– Battery of 12 V
– Working time minimally 1 h
– Input impedance > 10 MΩ
– Calibrational voltage 1 mV ± 2%

Every part of electrical medical device that comes into contact with patient's body has some risk of electrical shock caused by unsafe leakage currents. The electrical safety inspection involve testing of ground wire resistance, chassis leakage, patient leakage currents and mains on applied parts.

During the verification of ECG by etalon, range of measuring which must be controlled are next:

- Amplitude of voltage signal identified by ECG in mV is 0.5, 1.0, 1.5 or 2.0 mV
- Speed of beats in time frame of 1 min is: 30, 40, 60, 80, 90, 100, 120, 140, 150, 160, 180, 200, 210, 220, 240, 260, 270, 280, 300.
- Limits of allowed mistakes are:
- In case of measuring amplitude of voltage signal is ±5%
- In case of measuring speed of beats in time frame of 1 min is ±2%

Performance inspection is performed in order to determine measurement error of the device under test. If the error is in any of these cases bigger than maximum allowed error, ECG cannot be used and it must be serviced and verified again.

ECG device must have on visible place tile with accurately written label. These labels and marks should be written in language official for country.

Electrocardiographs must pass through procedure of examination and approval type, and before letting it into work, they must pass procedure of first verification and have certificates of verification. Examination of type is done based on documentation which producer or his agent must contribute along with application for approval. Documentation must have general, technical and other documentation and instructions for usage. First verification includes visual examination and is done by specific instruments. Maximum error allowed when done in regular verifications cannot extend maximum allowed by first verification. Periods of verification are defined by national regulations, e.g. [21].

6 Conclusion

ECG device is one of the biomedical instruments with wide history and enormous significance in everyday life of many people around the world. For centuries, scientists were trying to understand the principles and laws which this device would "obey". On the other side, some of them accidentally discovered secrets that were crucial for ECG development. From the 17th century, a dozen of different experiments were done by many scientists who, in that period, didn't know that their observations are going to be forerunner of today's most used device. Experiments started to improve in 18th century, so in 19th and 20th century development of ECG was done every day. Finally, in 1924, leading scientist acquired the most eminent award for his prominent achievement. That year was certainly one of the most important figures in history of science. However, it was not so easy to complete this task and produce consummate ECG. In any case, ECG device is real miracle. Its design nowadays seems simple for the engineers to understand, because of the technology rapidly moving forward. But, what is sure is that beginnings of this device were far more complex than anyone could possibly imagine. It is more than complex thing to invent any gadget without knowing how it will perform, will it be safe and is it going to serve as it should. Significance of the ECG is unimaginable. ECGs' usage is aimed at the heart—which enables us to be alive. When we think in that way, we can see of how great importance it is to human race. Heart diseases are number one in world right now, especially heart attack. With accurate ECG device, doctors can detect impending cardiac arrest and save life of many people. Also, whole spectrum of disorders or diseases can be seen, prevented and treated through the use of ECG and its features. It doesn't matter if we are talking about young athlete or elder, male or female, ECG is determining type of treatment, diet or any other important characteristic of persons' lifestyle. It's usage is not necessarily intended only for people who have heart issues. It should be used to prevent diseases by means of yearly control and tracking work of the heart. In a few years, it is expected for great improvements in the ECG systems to be achieved. Anyhow, device is used, and will be used always in everyday life because of:

– its features
– need for it
– no risk for patients, nor for doctors or nurses
– ensured diagnosis.

We can see that ECG has much of importance. Following that, people should get familiar with its basics; how does it work, why it is used and how to properly use it. Also, its availability should be very high, in order for wide population to always have access to it. For example, every pharmacy could have modern ECGs and people could test themselves in any moment. It would probably lower the need for seeing the doctor and at the same time would be useful to always have control of the heart.

There is never lack of the need to highlight significance of ECG devices. In a few years, it could be part of the daily routine, just like measuring body's temperature, arterial blood pressure or blood sugar concentration. ECG may become everyday habit.

Literature

1. Kramme R, Hoffmann K-P, Pozos R (eds) (2011) Springer handbook of medical technology. Springer, Berlin
2. Bronzino JD, Peterson DR (2015) The biomedical engineering handbook, vol 2, 4th edn. CRC Press, Boca Raton, FL
3. Einthoven W, History of ECG devices. [Online] Available at: https://www.nobelprize.org/nobel_prizes/medicine/laureates/1924/einthoven-bio.html. Last accessed June, 2017
4. AlGhatrif M, Lindsay J (2012) A brief review: history to understand fundamentals of electrocardiography. J Community Hosp Intern Med Perspect 2(1). doi:10.3402/jchimp.v2i1.14383

5. Ernstine AC, Levine SA (1928) A comparison of records taken with the Einthoven string galvanometer and the amplifier-type electrocardiograph. Am Heart J 4:725–731
6. Malmivuo J, Plonsey R (1995) Bioelectromagnetism—Principles and applications of bioelectric and biomagnetic fields. Oxford University Press, New York
7. Myerburg RJ et al (2008) Task force 2: training in electrocardiography, ambulatory electrocardiography, and exercise testing. J Am Coll Cardiol 51(3):348–354
8. Leitgeb N (2010) Safety of electromedical devices. Springer, Berlin. ISBN 978-3-211-99683-6
9. Buendía-Fuentes F et al. (2012) High-bandpass filters in electrocardiography: source of error in the interpretation of the ST segment. ISRN Cardiology 2012: 706217. PMC. Web. 18 July 2017
10. Northrop RB (2001) Noninvasive instrumentation and measurement in medical diagnosis. CRC Press, Boca Raton, FL. ISBN 9781420041200
11. Council Directive 93/42/EEC of 14 June 1993 concerning medical devices, [Online] Available at: http://eur-lex.europa.eu/LexUriServ/LexUriServ.do?uri=CELEX:31993L0042: EN:HTML
12. Regulatory framework, [Online] Available at: https://ec.europa.eu/growth/sectors/medical-devices/regulatory-framework_en
13. The new Regulations on medical devices, [Online] Available at: http://ec.europa.eu/growth/sectors/medical-devices/regulatory-framework/revision_hr
14. European Standards, [Online] Available at: http://ec.europa.eu/growth/single-market/european-standards/
15. IEC TC 62 Electrical equipment in medical practice, [Online] Available at: http://www.iec.ch/dyn/www/f?p=103:7:0::::FSP_ORG_ID,FSP_LANG_ID:1245,25
16. IEC 60601-1:2005 Medical electrical equipment—Part 1: General requirements for basic safety and essential performance, [Online] Available at: www.iec.org
17. IEC 60601-1-2:2014 Medical electrical equipment—Part 1-2: General requirements for basic safety and essential performance—Collateral Standard: Electromagnetic disturbances—Requirements and tests, [Online] Available at: www.iec.org
18. IEC/EN 60601-2-27 Medical electrical equipment—Part 2-27: Particular requirements for the basic safety and essential performance of electrocardiographic monitoring equipment, [Online] Available at: www.iec.org
19. IEC/EN 60601-2-47 Medical electrical equipment—Part 2-47: Particular requirements for the basic safety and essential performance of ambulatory electrocardiographic systems, [Online] Available at: www.iec.org
20. IEC/EN 60601-2-51 Medical electrical equipment—Part 2-51: Particular requirements for safety, including essential performance, of recording and analysing single channel and multichannel electrocardiographs, [Online] Available at: www.iec.org
21. Rule book on metrology and technical requirements for electrocardiographs.http://www.met. gov.ba/dokumenti/PRAVILNIK%20O%20MJERITELJSKIM%20I%20TEHNICKIM% 20ZAHTJEVIMA%20ZA%20ELEKTROKARDIOGRAF_BOSANSKI.pdf
22. Webster JG (2011) Medical instrumentation application and design, 4th edn. Wiley, New York

Inspection and Testing of Noninvasive Blood Pressure Measuring Devices

Igor Lacković

1 Introduction

The measurement of blood pressure is important in the diagnosis and monitoring of a wide range of clinical conditions. The blood pressure in the circulation is principally due to the pumping action of the heart and other determinants including peripheral vascular resistance, the blood volume and viscosity. The pumping action of the heart generates pulsatile blood flow, which is conducted into the arteries, across the micro-circulation and back via the venous system to the heart. Blood pressure usually refers to the arterial blood pressure in the systemic circulation. Arterial blood pressure is the force blood exerts per unit area on the walls of the arteries as the heart pumps it through the arterial tree. It is one of the vital signs, along with heart rate, oxygen saturation, respiratory rate and body temperature. Blood pressure is usually expressed in terms of the systolic pressure (maximum during one heart beat) over diastolic pressure (minimum in between two heart beats). Illustration of idealized arterial pressure waveform with indication of characteristic point pressures is given in Fig. 1. For a more detailed introduction to cardiovascular anatomy and physiology refer to [1].

Normal blood pressure at rest for adults is within the range of 100–140 mm of mercury (mmHg) systolic and 60–90 mmHg diastolic. Hypertension, or high blood pressure is present if the resting blood pressure is persistently at or above 140/90 mmHg for most adults. Table 1 gives one of the most widely used classifications of blood pressure for adults [2]. As of 2015, approximately one billion adults or ∼22% of the population of the world have hypertension [3]. Arterial hypertension is a major risk factor for heart disease and could lead to severe organ damage. It is often called a "silent killer", because there are usually no warning

I. Lacković (✉)
Faculty of Electrical Engineering and Computing,
University of Zagreb, Unska 3, 10000 Zagreb, Croatia
e-mail: igor.lackovic@fer.hr

© Springer Nature Singapore Pte Ltd. 2018
A. Badnjević et al. (eds.), *Inspection of Medical Devices*, Series in Biomedical Engineering, https://doi.org/10.1007/978-981-10-6650-4_5

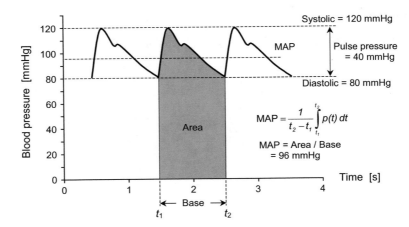

Fig. 1 Idealized arterial pressure waveform. When the left ventricle ejects blood into the aorta, the aortic pressure rises. The maximal aortic pressure following ejection is termed the systolic pressure (Psys). As the left ventricle is relaxing and refilling, the pressure in the aorta falls. The lowest pressure in the aorta, which occurs just before the ventricle ejects blood into the aorta, is termed the diastolic pressure (Pdia). MAP indicates mean arterial pressure. MAP $\approx$ 2/3(Pdia) + 1/3(Psys)

Table 1 Classification of blood pressure for adults (JNC7) [2]

Category	Systolic pressure (mmHg[a])	Diastolic pressure (mmHg)
Normal	90–119	60–79
High normal (Prehypertension)	120–139	80–89
Stage 1 hypertension	140–159	90–99
Stage 2 hypertension	160–179	100–109
Stage 3 hypertension (emergency)	≥ 180	≥ 110
Isolated systolic hypertension	≥ 140	<90

[a]A millimeter of mercury is a manometric unit of pressure, formerly defined as the extra pressure generated by a column of mercury one millimetre high and now defined as precisely 133.322387415 pascals. Although not an SI unit, the millimeter of mercury is still routinely used in medicine, meteorology and some other scientific fields

symptoms before hypertension strikes a person in the form of a stroke, heart attack, heart failure, eye problems (hypertensive retinopathy) or kidney disease. Therefore it is important to measure blood pressure regularly especially if any of risk factors (family history, smoking, obesity, high sodium intake, stress, etc.) is present.

Auscultatory method using aneroid or mercury sphygmomanometers is commonly used for manual blood pressure measurements. For automated noninvasive blood pressure monitors oscillometric method has become the de facto standard. Public awareness of risks associated with high blood pressure and the availability of embedded microcontroller-based systems have resulted in the development of numerous automated devices for noninvasive blood pressure measurement and their widespread use not only in medical facilities, but also in homes and public places. The only way to determine whether someone has high blood pressure is to have it checked regularly. The key to blood pressure control is "good blood pressure measurement". From the engineering point of view "good" means accurate and reliable. Both medical professionals and general public users require accurate, safe and reliable blood pressure measuring. A standardized set of recommendations for blood pressure measurement in humans, that, if followed, should lead to accurate estimation of blood pressure, are summarized in [4].

2   Survey of Methods for Blood Pressure Measurement

Methods for arterial pressure measurement are usually classified into direct and indirect methods [5]. The first being invasive and the only that measure the "true" pressure. All other methods belong to the group of indirect methods since the pressure is measured noninvasively from outside the body.

2.1   Invasive Method

Invasive methods imply the insertion of an arterial cannula into a suitable artery and then displaying the measured pressure waveform on a monitor. The arterial cannula is connected to tubing filled with saline, which acts as a coupling medium between the blood in the artery and the external pressure transducer. The liquid within the tubing is in contact with a diaphragm that moves in response to the transmitted pressure wave. The movement is converted to an electrical signal by a transducer. In that way the complete arterial pressure waveform is measured and it is easy to determine systolic and diastolic pressure on a beat-to-beat basis.

Invasive method is the gold standard of blood pressure measurement giving accurate beat-to-beat information. Due caution is required regarding the frequency response of the system (i.e. damping, resonant frequency, etc.) as it affects the accuracy of intraarterial pressure monitoring. Arterial pressure waveform is a complex waveform composed of many individual sine waves. It is therefore important that the natural frequency of the measuring system (the catheter and column of saline etc.) does not correspond to any of the frequencies of the arterial pressure waveform. This is achieved by making sure that the natural frequency of the measuring system is raised high enough i.e. above any of the frequencies of the arterial pressure waveform. Therefore arterial catheter must be short and with the maximum gauge possible; column of saline must be as short as possible; the catheter and tubing must be stiff walled; the transducer diaphragm must be as rigid as possible. Also, a potential source of error may be the incorrect positioning of the catheter or the pressure transducer positioned at the different level to the patient's heart. The drawbacks of direct methods are that they are invasive and uncomfortable for patients and in long-term use could lead to risks associated with infection, air embolism or thrombosis. Therefore invasive measurement of blood pressure is performed only in clinical environment in patients who are likely to display sudden changes in blood pressure (e.g. vascular surgery), in whom close control of blood pressure is required (e.g. head injured patients), or in patients receiving drugs to maintain the blood pressure.

2.2   Noninvasive Methods

Noninvasive measurement of arterial pressure is based on the detection of certain characteristic physical phenomena that can be registered at the surface of the body and correlating these phenomena to the arterial pressure. In the majority of noninvasive methods an occlusive cuff is used to obstruct the blood flow normally in the brachial artery. Then during cuff deflation, occurring phenomena are being recorded (e.g. Korotkoff sounds, cuff pressure oscillations, etc.). In some methods systolic and diastolic pressure can be determined not only during cuff deflation but also during cuff inflation. The process of cuff deflation (or inflation) can be continuous or incremental (stepwise).

Noninvasive methods may be classified according to different criteria. A widely used criterion is whether the method enables continuous or intermittent measurement. Main feature of intermittent methods is that systolic and diastolic pressures are obtained during the time interval that encompasses many heartbeats. Continuous methods provide either absolute, continuous pressure waveform similar to intraarterial recording (from which characteristic pressures are easily determined) or give only the systolic and diastolic pressure on a beat-to-beat basis. The main advantage of continuous methods is their ability to track beat-to-beat variations of blood pressure. The principal disadvantage of all continuous noninvasive methods is a relatively large measurement error that depends on the calibration method being used. Continuous noninvasive methods have great importance in long-term physiological monitoring, e.g. for studies of sleep disorders or nocturnal hypertension, or in polygraphic recordings.

Noninvasive methods can also be classified upon the use of cuff. We can distinguish cuffless methods, methods that use partially inflated cuff and methods that use a cuff for complete occlusion of the arterial blood flow (cuff pressure is raised above systolic pressure). If occlusive cuff is used at least ~2–5 min are required between successive measurements to allow for restoration of normal blood flow.

Based on the physical principle noninvasive methods for blood pressure measurement are divided into one of the following categories [5, 6]:

- Auscultatory method (Riva-Rocci method, Korotkoff method)
- Oscillometric method
- Palpatory method
- Ultrasound method
- Pulse-wave velocity method (Transit-time method)
- Vascular unloading method
- Arterial tonometry

Among these, the most widely used are the auscultatory method and the oscillometric method. Other methods are not routinely used. They are either used only in research settings or were used in the past and have been replaced by other methods.

Some characteristics of noninvasive methods for blood pressure measurement are disclosed in Table 2.

It is important to stress that each method has its own algorithms (criteria) to determine characteristic pressures that will be discussed in the forthcoming sections. Therefore the accuracy of any particular method cannot be established without knowing the characteristics of the algorithms used to identify systolic and diastolic pressure. That is especially important for the oscillometric method since numerous algorithms have been developed over the years, and also for the auscultatory method due to the disagreement whether Phase IV or Phase V of Korotkoff sounds should be used as the indicator of diastolic pressure.

Table 2 Characteristics of noninvasive blood pressure measurement methods

Method	Intermittent/Continuous	Cuff/Cuffless
Auscultatory method	Intermittent	Occlusive cuff
Oscillometric method	Intermittent	Occlusive cuff
Palpatory method	Intermittent	Occlusive cuff
Ultrasound method	Intermittent	Occlusive cuff
Pulse-wave velocity method	Continuous	Cuffless
Vascular unloading method	Continuous	Partially inflated cuff
Arterial tonometry	Continuous	Cuffless although the artery is partially occluded

2.2.1 Auscultatory Method

The auscultatory method (Riva-Rocci method, Korotkoff method) is still the most widely used noninvasive method for blood pressure measurement. For routine arterial pressure measurement, physicians prefer the manual auscultatory method. Automatic devices based on the use of one or several microphones were also developed.

Occlusive cuff, typically 12 cm wide is wrapped around the upper arm and rapidly inflated to the supra systolic pressure in order to completely stop the blood flow distal to the cuff. To avoid erroneous results, due caution is required to the choice of proper cuff width (standard 12 cm cuffs are usually inappropriate for obese, or generally always when subject's arm circumference considerably deviates from average population), as well as to the proper cuff positioning and the degree of tightening. The cuff deflation rate should be around 3 mmHg per beat or approximately 3 mmHg/s. As the cuff pressure is decreased, a stethoscope placed over the artery distal to the cuff will detect a sequence of sounds (the Korotkoff sounds) that suddenly appear (Phase I), change in character and intensity (Phase II, Phase III, Phase IV) and than gradually disappear (Phase V), Fig. 2. The appearance of

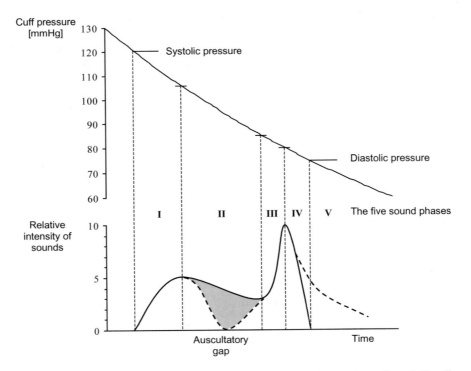

Fig. 2 Relative intensity of Korotkoff sounds, along with the estimate of systolic and diastolic pressure

sounds (Phase I) is taken as the indicator of systolic pressure, and the disappearance of sounds (Phase V) is usually used as the indicator of diastolic pressure. Mean arterial pressure cannot be determined, and if needed, a suitable empirical formula is used to calculate it from systolic and diastolic pressure.

The phenomenon known as auscultatory gap, which is characterized by disappearance and then reappearance of Korotkoff sounds, can lead to an erroneous indication of diastolic pressure (see Fig. 2). To avoid errors due to auscultatory gap cuff pressure should be allowed to fall below the gap, where sounds return. Rapid cuff inflation and increased interval between successive measurements can also help, since auscultatory gap frequently occurs due to high venous pressure distal to the cuff.

In some subjects, during cuff deflation, the sounds do not disappear after Phase IV (Phase V does not exists). Instead, they persist to well below diastolic pressure (see Fig. 2). In that case the beginning of Phase IV is recommended as the indicator of diastolic pressure.

The origin of Korotkoff sounds is in two physically different phenomena—rapid movement of the arterial wall and turbulent flow of blood through partially opened artery [7]. Clearly, the contribution of each is not equal and dominant mechanism changes during phases.

Auscultatory method can be automatized in a way that cuff is automatically inflated and slowly deflated and that systolic and diastolic pressure are measured electronically. In that case instead of stethoscope, one or more microphones are used, and an electronic system for signal processing is needed. The limitation of that approach is high sensitivity to external noise that can lead to large errors. Techniques to reduce the effects of artifacts and external noise include the application of more microphones and ECG-gating or oscillometric-pulse-gating (OP-gating) techniques. Identification of Korotkoff phases can be performed either in time or in frequency domain. Frequency spectrum of Korotkoff sounds covers frequency range from 20 Hz to 300 Hz, but the majority of energy content lies below 50 Hz. The Korotkoff sounds represent only the audible portion of a broader range of arterial vibrations that also spread to inaudible range.

Some of the most important factors that influence the accuracy of auscultatory method are: cuff width, observer's ability to identify the phases, choice of Phase IV or Phase V as indicator of diastolic pressure, manometer accuracy, cuff deflation rate, cuff location (influence of hydrostatic pressure), etc.

Comparison of auscultatory method with direct intraarterial recording performed by London and London, showed that Korotkoff method underestimates systolic pressure by 5–20 mmHg and overestimates diastolic pressure by 12–20 mmHg [8]. Average error estimates were −2 mmHg for systolic and +4 to +10 mmHg for diastolic pressure. Another study showed similar results [9].

In summary, the average error of blood pressure measured by auscultatory method is around 10 mmHg when compared to intraarterial recording. Auscultatory method tends to underestimate systolic pressure and to overestimate diastolic pressure. Measurement error is larger for diastolic than for systolic pressure.

2.2.2 Oscillometric Method

The oscillometric method is based on characteristic physical phenomenon—cuff pressure oscillations (oscillometric pulses) generated by the pulsatile displacement of the artery that occur during cuff deflation from supra systolic to sub diastolic pressures. With decreasing cuff pressure the amplitude of the oscillations increases at first, reaches its maximum and then begins to decrease, Fig. 3. These phenomena are used to calculate characteristic pressures [6, 10, 11].

Mean arterial pressure is easiest to determine—it corresponds to oscillations' maximum. Systolic and diastolic pressure are determined indirectly, since oscillometric method provides clear indicator only for mean arterial pressure. There are two types of criteria used to determine systolic and diastolic pressure: height-based criteria (amplitude ratio approach) and slope-based criteria (derivative oscillometry).

Criteria for systolic, diastolic and mean arterial pressure determination can be graphically demonstrated when the oscillations' envelope is plotted against the corresponding baseline cuff pressure, Fig. 4.

In the height-based approach (amplitude-ratio approach), the systolic pressure is determined as the baseline cuff pressure that is greater than the mean arterial pressure and at which the ratio of the oscillometric pulse amplitude A_s over the maximum pulse amplitude A_{max} is equal to a certain predetermined value—the systolic ratio. Similarly, the diastolic pressure is determined as the baseline cuff

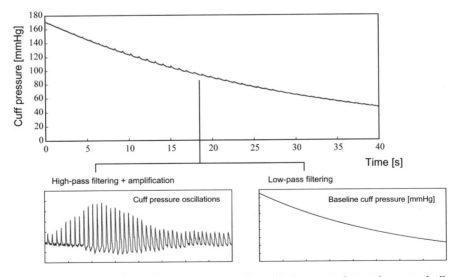

Fig. 3 Cuff pressure signal (top) contains small oscillations superimposed on gradually decreasing pressure. After high pass filtering and amplification dynamics of cuff pressure oscillations is clearly observed. Low-pass filtering removes oscillometric pulses to obtain baseline cuff pressure (bottom)

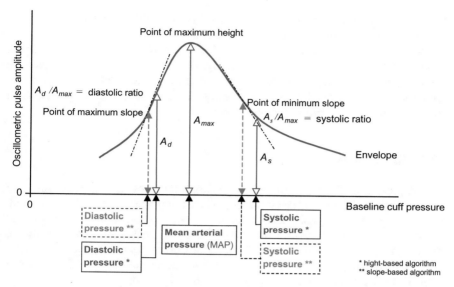

Fig. 4 Envelope of cuff pressure oscillations (oscillometric pulses) with the illustration of two algorithms for detection of systolic and diastolic pressure. One asterisk—height based approach (amplitude ratio approach), two asterisk—slope based approach

pressure that is lower than the mean arterial pressure and at which the ratio of the oscillometric pulse amplitude A_d over the maximum pulse amplitude A_{max} is equal to another predetermined value—the diastolic ratio. These ratios depend on cuff compliance, cuff deflation rate, etc. and have to be determined empirically on a population of subjects. Systolic ratio is usually around 0.5 and diastolic ratio around 0.7, although literature values range from 0.4 to 0.75 for systolic and from 0.6 to 0.86 for diastolic ratio.

In the slope-based approach the baseline cuff pressure at which the pulse amplitude increases rapidly is taken as systolic pressure, while that at which the amplitude decreases rapidly is taken as the diastolic pressure. Mathematically, these two points may be defined as the points at which the slope of the envelope is maximum or minimum (inflection points of the envelope). Equivalently, time derivative of the oscillations' envelope can be used to determine characteristic pressures. Therefore, slope-based approach is also known as derivative oscillometry. Mean arterial pressure can be easily identified as the point where the derivative passes through zero. The baseline cuff pressure at which the derivative reaches maximum is taken as the systolic pressure, while the baseline cuff pressure at which the derivative reaches minimum is taken as the diastolic pressure.

Different variations of oscillometric method have been reported both regarding the algorithms, the application of more than one cuff etc., but many technical details used in commercial oscillometric devices are proprietary and patent protected. Very similar to oscillometric method is the volume-oscillometric method. The only fundamental difference is that volume-oscillometric method is based on artery volume oscillations. Consequently photoplethysmographic or similar volumetric sensor is used for detecting arterial volume changes.

The key point to understanding limitation of oscillometric method is that it provides only mean arterial pressure directly. Both systolic and diastolic pressure are calculated using empirical criteria. Also, motion artifacts, presence of arrhythmias, increased arterial stiffness due to aging, etc. can lead to envelope distortion and erroneous measurement of blood pressure.

2.2.3 Other Methods

The **palpatory method** (return-to-flow method) can provide a good estimate of systolic pressure [5, 6]. During cuff deflation, an artery distal to the cuff is palpated. The appearance of palpable beats indicates systolic pressure. Instead of manual palpation, an electronic sensor can be used. Electronic palpation is more sensitive than manual and enables more accurate determination of systolic pressure. Usually electronic palpatory methods use arm cuff and transcutaneous doppler flowmeter or other plethysmographic sensor (typically photoplethysmographic) that is placed at the wrist or finger. As long as the cuff pressure is above the systolic pressure, blood flow distal to the cuff is obstructed and the sensor cannot detect any changes of blood volume. When the cuff pressure falls below the systolic pressure, blood starts to flow through the partially opened artery to the distal part of the limb. That moment can be detected by a distal sensor as appearance of pulses and indicates systolic pressure. However, diastolic pressure cannot be determined by palpatory method since there is no indicator in the distal pulse when cuff pressure passes through diastolic pressure.

The **ultrasound method** is based on the ultrasonic detection of the arterial wall motion during cuff deflation. Two piezoelectric transducers, one that serves as transmitter and generates ultrasound waves, and the other that receives the reflected waves are embedded in the distal part of the occlusive cuff. Systolic and diastolic pressure are determined from the frequency shift of the reflected waves to the transmitted waves due to the Doppler effect [5]. As long as the cuff pressure in greater than systolic pressure, artery is fully occluded and the transmitted waves reflect without the change of frequency. When the cuff pressure falls below the systolic pressure, the artery wall moves rapidly what is manifested with the appearance of high frequency and immediately after by the low frequency due to the closing of the artery. As cuff pressure decreases, time between opening and closing increases causing the increase of time interval between frequency shifts. Finally closing and next opening coincide and frequency shift disappears. That indicates the diastolic pressure. Although ultrasound method has a relatively small error, complicated measuring system and the need for contact medium between the transducers and the skin renders this method unsuitable for clinical practice or home use.

The **pulse-wave velocity method** proceeds from the fact that pulse-wave velocity (the velocity of pressure wave along the arterial tree) increases with the increase of arterial pressure [6]. Theoretically, this relationship can be used to derive arterial pressure waveform, without a need for a cuff, which if calibrated against measurements made by a reference method, yields absolute continuous arterial pressure waveform. However, measurement of pulse-wave velocity is indirect—pulse wave velocity is computed from the measured pulse-transit time between two different sites of the arterial tree during the same cardiac cycle and the estimated distance between these two sites. Although the pulse-wave velocity method is theoretically capable of measuring continuous arterial pressure waveform, it typically provides only beat-to-beat point pressure. To obtain continuous pressure waveform multiple measurements of the pulse–transit time during the same cardiac cycle would have to be made for each and every heartbeat. That problem is still not solved. In order to measure pulse-transit time, two photoplethysmographic sensors are used. Typical peripheral measuring sites include forehead, finger and wrist. Instead of using two photoplethysmographic sensors it is also possible to measure the time between the occurrence of the R-wave of ECG and the arrival of the pulse wave at the peripheral site that is detected by a photoplethysmographic sensor [12]. Modifications of pulse-wave velocity method include the application of occlusive cuff. In that case the pulse-wave velocity depends on the degree of artery occlusion. When the cuff pressure is just below the systolic pressure, the pulse-wave velocity is low, and consequently the measured pulse transit time to the distal sensor is large. As the cuff pressure decreases the transit time gradually becomes shorter reaching a steady state near the cuff pressure value that is near the diastolic pressure. This modified pulse-wave velocity method provides only intermittent pressure measurement.

The **vascular unloading method** (also known as Penaz method, or FINAPRES method) belongs to the group of continuous measurement methods and is usually performed at a finger. The method is based on the following theory [13]. If an external pressure applied to the artery is continuously changed as to be equal to the arterial pressure, the artery will be continuously unloaded meaning that the transmural pressure (difference between intraarterial and external pressure) across the artery wall is zero. In that state the artery has maximal compliance what corresponds to the maximal arterial volume oscillations. These volume oscillations are detected by a photoplethysmographic transducer. By applying the mechanical control loop, cuff pressure is continuously adjusted to support the continuous unloading of the arterial wall. In that way the cuff pressure continually follows and is equal to the intraarterial pressure meaning that complete pressure waveform is obtained and systolic and diastolic pressure can be easily determined.

The method of **arterial tonometry** is based on the fact that if a superficial artery, close to an underlying bone is partially flattened or applanated with a flat rigid surface, and kept in that state, the force exerted to the surface is proportional to the arterial pressure [6, 14]. This relationship can be used to derive the arterial pressure waveform, which if calibrated yields absolute continuous arterial pressure. Tonometer is usually placed at the wrist and radial artery pressure is measured. The instrument consists of a pressure transducer (an array of piezoresistive sensors), an electropneumatic system to applanate the artery and an electromechanical positioning system. However, these devices are difficult to position, and maintaining the proper contact is a challenge [14].

3 Blood Pressure Measurement Equipment

Devices for noninvasive blood pressure measurement, usually known as NIBP monitors (NIBP stands for noninvasive blood pressure), can be classified into three groups. One of them is the group of ambulatory monitors for 24 h recording of arterial pressure during normal activities, similar to ambulatory ECG. These are designed to record the patient's blood pressure at pre-defined intervals over a 24-h period during normal activities and store the data for future analysis. These devices help physicians to diagnose blood pressure disorders and to manage and optimize anti-hypertension therapy. Ambulatory devices are also important to assess the prevalence of "white-coat" hypertension (elevated blood pressure when measured in the physicians' office due to emotional excitement or fear). The second group comprises of bedside and transport monitors. The blood pressure measurement module is usually a part of multiparameter physiological monitor that enables measurement of ECG, SpO_2, respiration, body temperature, etc. These make repetitive measurements at set time intervals and often incorporate vital sign parameter alarms. They are designed for bed-side monitoring in a clinical environment and are an expensive option. The third group of blood pressure measuring devices is the largest one comprised of the so-called self-taking NIBP monitors. The characteristic of this group is that these devices are intended for routine measurement at home, office or in public places either by a subject himself or by a physician. These devices are usually very simple and are widely available for an attractive price.

According to the way the systolic and diastolic pressure are determined and the way the cuff pressure is controlled three categories of NIBP monitors can be distinguished: non-automatic devices (manual cuff inflation, observer determines characteristic pressures), semiautomatic devices (manual cuff inflation, automatic determination of characteristic pressures) and fully automatic devices (automatic cuff inflation, automatic pressure determination). Typical non-automatic device is mercury sphygmomanometer, typical semi-automatic devices are some older oscillometric blood pressure monitors with manual cuff pump, and the example of the automatic devices are the majority of oscillometric NIBP monitors available nowadays at the market. Combined auscultatory-oscillometric monitors are also available.

3.1 Devices for Use with the Manual Auscultatory Method

The auscultatory method relies on inflating an upper arm cuff to occlude the brachial artery and then listening to the Korotkoff sounds through a stethoscope whilst the cuff is slowly deflated. The patient's systolic (phase I) and diastolic blood pressure (phase V) is recorded from the reading on the sphygmomanometer. Devices for use with the manual auscultatory method, Fig. 5, [15]:

- **Mercury sphygmomanometer**
  This includes a mercury manometer, an upper arm cuff and a hand inflation bulb with a pressure control valve; requires the use of a stethoscope to listen to the Korotkoff sounds. The mercury sphygmomanometer has always been regarded as the gold standard for clinical measurement of blood pressure. In principle, the simplicity of the design means that there is negligible difference in the accuracy of different brands and that there is less to go wrong with mercury sphygmomanometers than with any other type of manometer.
- **Aneroid sphygmomanometer**
  As above, with an aneroid gauge replacing the mercury manometer. In these devices, the pressure is registered by a mechanical system of metal bellows that expands as the cuff pressure increases and a series of levers that register the pressure on a circular scale. The aneroid gauge may be wall or desk mounted or attached to the hand bulb.
- **Electronic sphygmomanometer**
  As above, with a pressure sensor and electronic display replacing the mercury manometer. Blood pressure is taken in the same way as with a mercury or aneroid device, by an observer using a stethoscope and listening for the Korotkoff sounds. The cuff pressure can be displayed as a simulated mercury column, as a digital readout, or as a simulated aneroid display. Battery powered.

3.2 Automated Devices, Generally Using the Oscillometric Method

The majority of non-invasive automated blood pressure measuring devices currently available use the oscillometric method. The oscillometric method relies on detection of variations in pressure oscillations due to arterial wall movement beneath an occluding cuff. Empirically derived algorithms are employed, which

Fig. 5 Devices for use with the manual auscultatory method (from left to right: mercury sphygmomanometer; aneroid sphygmomanometers (hand-held model and wall-mounted model); electronic sphygmomanometer)

calculate systolic, mean arterial and diastolic blood pressure. Manufacturers develop their own algorithms by studying a population group and may have validated the stated accuracy by performing a clinical trial in accordance with one of the standards (AAMI/ANSI SP10, BHS, DIN, etc.). Automated devices, generally using the oscillometric method, Fig. 6, [15]:

- **Automated (spot-check) device**
  This includes an electronic monitor with a pressure sensor, a digital display and an upper arm cuff. An electrically-driven pump raises the pressure in the cuff. When started, the device automatically inflates the cuff to the appropriate level (usually about 30 mmHg above an estimated systolic reading), then deflates the cuff and displays the systolic and diastolic values. Some devices may have a user-adjustable set inflation pressure. The majority calculate these values from data obtained during the deflation cycle, but there are some that use data from the inflation cycle. The pulse rate may also be displayed. These devices may also have a memory which stores the last measurement and previous readings. Battery powered.
- **Wrist device**
  This includes an electronic monitor with a pressure sensor, an electrically-driven pump attached to a wrist cuff. Function is similar to the automated (spot-check) device above. Battery powered.
- **Finger device**
  This includes an electronic monitor and a finger cuff, or the device itself may be attached to the finger. Battery powered. Uses oscillometric, pulse-wave or plethysmographic methods for measurement.
- **Spot-check non-invasive blood pressure (NIBP) monitor**
  This is a more sophisticated version of the automated device above and is designed for routine clinical assessment. There may be an option to measure additional vital signs, such as oxygen saturation in the finger pulse (SpO2) and body temperature. Mains and battery powered.
- **Automatic-cycling non-invasive blood pressure (NIBP) monitor**
  This is similar to the spot-check NIBP monitor, but with the addition of an automatic-cycling facility to record a patient's blood pressure at set time intervals. These are designed for bed-side monitoring in a clinical environment where repetitive monitoring of patients and an alarm function is required. These devices may incorporate the ability to measure additional vital signs. The alarm limits can usually be set to alert nursing staff when one or more of the measured patient parameters exceed the pre-set limits. Mains and battery powered.
- **Multi-parameter patient monitors**
  These are designed for use in critical care wards and operating theatres and monitor a range of vital signs including blood pressure. May be possible to communicate with a Central Monitoring Station via Ethernet or Wi-Fi.
- **Ambulatory blood pressure monitor**
  This includes an upper arm cuff and an electronic monitor with a pressure sensor and an electrically-driven pump that attaches to the patient's belt. The unit is

programmed to record the patient's blood pressure at pre-defined intervals over a 24-h period during normal activities and stores the data for future analysis. Battery powered. Uses electronic auscultatory and oscillometric methods.

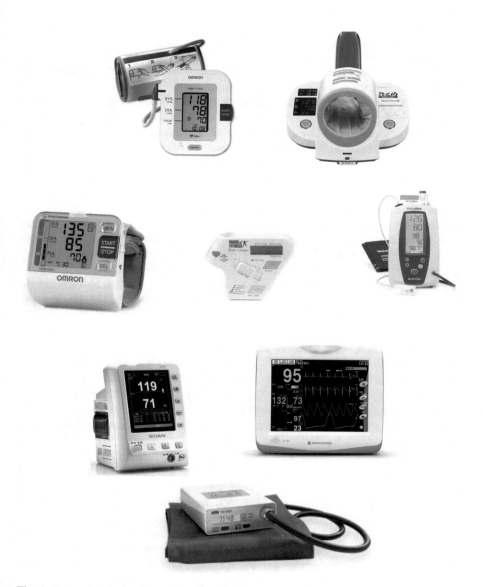

Fig. 6 Automated devices for noninvasive blood pressure measurement (from top left to bottom right: automated spot check devices, wrist device, finger device, spot check NIBP monitor, cycling NIBP monitor, multiparameter patient monitor, ambulatory blood pressure monitor)

4 Sources of Error, Type of Hazards and Other Issues

Sources of error for blood pressure measuring devices could be manometer related, cuff related, patient related and observer related. For automated devices errors could be also algorithm related.

As stated previously, the mercury sphygmomanometer has always been regarded as the gold standard for clinical measurement of blood pressure, but nowadays these are being removed from clinical practice because of environmental concerns about mercury contamination. In principle, the simplicity of the design means that there is negligible difference in the accuracy of different brands and that there is less to go wrong with mercury sphygmomanometers than with any other type of manometer. Even in most recent International protocol for the validation of blood pressure measuring devices by the European Society of Hypertension (from 2010) mercury sphygmomanometers are still used as reference standards. However, one hospital survey found that 21% of devices had technical problems that would limit their accuracy [16]. In another study just under 500 mercury sphygmomanometers and their associated cuffs were examined and more than half had serious problems that would have rendered them inaccurate in measuring blood pressure [17]. In an aneroid sphygmomanometer, aneroid gauge is used replacing the mercury manometer. This type of system does not necessarily maintain its stability over time, particularly if handled roughly (shocks, drops, etc.). Aneroid sphygmomanometers therefore are inherently less accurate than mercury sphygmomanometers and require calibrating at regular intervals. Recent developments in the design of aneroid devices may make them less susceptible to mechanical damage when dropped. Wall-mounted devices are less susceptible to mechanical shocks and are generally more accurate than mobile devices [18]. Surveys conducted in hospitals have examined the accuracy of the aneroid devices and have shown significant inaccuracies ranging from 1% [18, 19] to 44% [16]. The accuracy of the manometers varies greatly from one manufacturer to another. Some studies have focused on the accuracy of the pressure registering system and have identified that small dials used in many of the devices limit their accuracy. Another survey of the accuracy of the absolute static pressure scale of aneroid, mercury and automated sphygmomanometers in clinical use in primary care revealed that 17.9% of all surveyed devices gave errors exceeding the ±3 mmHg threshold [20].

Apart from the before mentioned issues related to the accuracy of manometer or pressure registering system, it is also important to recognize that all manual techniques may suffer from observer bias including differences of auditory acuity between observers [15]. Also digit preference is common, with observers recording a disproportionate number of readings ending in five or zero. The observer may also be influenced by the knowledge that they have of the patient (i.e. earlier readings, expected effect of drug therapy, gender, age, race and weight). However, formal training in blood pressure measurement can improve this situation.

Automated methods are also not error free. Users should be aware that for patients experiencing muscle tremors, abnormal heart rhythms, weak pulse or very low blood pressure due to shock, some automated blood pressure devices may fail to obtain a reading and will either indicate an error code or give unreliable results [15]. It is important to recognize the limitations of automated oscillometric devices in certain groups (e.g. those with cardiac arrhythmias, pre-eclampsia and some vascular diseases).

Incorrect cuff size (i.e. cuff width) is also a source of error for both manual sphygmomanometers and automated blood pressure measuring devices. An under-sized cuff tends to over-estimate blood pressure, while an over-sized cuff may under-estimate. This is especially critical when measuring blood pressure in children or obese. Incorrect cuff placement can also be a source of error. The cuff should be placed on the arm with the centre of the bladder over the brachial artery. The optimum bladder size is considered to be: width 40% of limb circumference, length 80–100% of limb circumference at the centre of the range for each cuff size [15].

5 Standards and Protocols for Evaluation of Blood Pressure Measuring Devices

From the 1990s different national and international associations and standardization organizations (ANSI/AAMI—American National Standards Institute/Association for the Advancement of Medical Instrumentation, BHS—British Hypertension Society, European Society for Hypertension, ISO—International Organization for Standardization, IEC—International Electrotechnical Comission, etc.) have set up standards and testing protocols for evaluation of blood pressure measuring devices [21–39]. Some of these standards and protocols are shown in Tables 3 and 4. Apart from those specific standards for blood pressure measuring devices general standards for medical electrical equipment safety and performance (i.e. IEC 60601-1) also apply.

Since automatic noninvasive blood pressure monitors are widely used both in clinical environments and homes, and are probably the most interesting to the readers, some details on the evaluation of these devices according to AAMI (1992), BHS (1993) and DIN (1995) protocol are listed in Table 5.

Considering the number and characteristics of subjects, there is not much difference between them. Minimum 85 subjects is required if manual auscultatory method is used as reference and minimum 15 subjects if direct intraarterial recording is performed. Also, sex and age is mainly left to chance. The AAMI (1992) gives special hints to some groups of patients (elderly, diabetic persons) while both AAMI and BHS (1993) protocol require testing of the device on a prescribed number of persons with particular blood pressure values. Different well-trained observers should perform comparative measurements. There is also much agreement between the three protocols concerning the allowable error of the

Table 3 Some standards and protocols for evaluation of blood pressure measuring devices

Blood pressure measuring device	Standard or protocol
Noninvasive non-automated blood pressure measuring devices	ANSI/AAMI SP9 (1994), Non-automated sphygmomanometers
	ISO 81060-1:2007 Non-invasive sphygmomanometers —Part 1: Requirements and test methods for non-automated measurement type
Noninvasive automated blood pressure (NIBP) monitors	ANSI/AAMI SP10: (1992) for electronic or automated sphygmomanometers and its revisions
	IEC 80601-2-30:2009 + AMD1:2013 CSV Consolidated version Medical electrical equipment— Part 2-30: Particular requirements for the basic safety and essential performance of automated non-invasive sphygmomanometers
	The British Hypertension Society (BHS) protocol (1993) for the evaluation of blood pressure measuring devices and its revisions
	DIN 58130 (1995) Nichtinvasive Blutdruckmeßgeräte— Klinische Prüfung which has been replaced by DIN EN 1060-4:2004-12, and more recent DIN EN ISO 81060-2:2014-10
Invasive blood pressure measuring devices	IEC 60601-2-34:2011 Medical electrical equipment— Part 2-34: Particular requirements for the basic safety and essential performance of invasive blood pressure monitoring equipment
	ANSI/AAMI BP22:1994/(R)2011 Blood pressure transducers

AAMI—Association for the Advancement of Medical Instrumentation
ANSI—American National Standards Institute
BHS—British Hypertension Society
DIN—Deutsches Institit für Normung
IEC—International Electrotechnical Comission
ISO—International Organization for Standardization

systolic and diastolic pressure. The AAMI (1992) and DIN (1995) limits are identical, while BHS (1993) protocol requires different report of results.

Table 6 gives the comparison of the British Hypertension Society Protocol BHS (1993), the International Protocol of the European Society of Hypertension ESH IP (2002) and revised International protocol ESH IP (2010).

Tables 5 and 6 should not be used without knowledge of the full protocols. They are intended solely as a quick comparison between protocols.

EN 1060-1 [35] states that for both increasing and decreasing pressure, the maximum error for the measurement of the cuff pressure at any point of the scale range shall be ±3 mmHg. This applies to the 'static calibration' of the cuff pressure in manual and automated devices. This belongs to legal metrology inspection (more details in Sect. 6).

ISO 81060-2 [26] states for systolic and diastolic blood pressures, the mean value of the differences of the determinations shall be within or equal to ±5 mmHg,

Table 4 Standards, validation protocols, recommendations

Standard/validation protocol	References
Sphygmomanometer Standards	European Standard EN 1060-1:1996. Specification for Non-invasive sphygmomanometers. Part I. General requirements. 1995. European Commission for Standardisation. Rue de Stassart 36, B-1050 Brussels [35]
	European Standard EN 1060-2:1996. Specification for Non-invasive sphygmomanometers. Part 2. Supplementary requirements for mechanical sphygmomanometers. 1995. European Commission for Standardisation. Rue de Stassart 36, B-1050 Brussels [36]
	European Standard EN 1060-3: 1997. Non-invasive sphygmomanometers. Part 3. Supplementary requirements for electro-mechanical blood pressure measuring systems. European Committee for Standardization. Rue de Stassart 36, B-1050, Brussels [37]
International Protocol of the European Society of Hypertension	O'Brien E, Atkins N, Stergiou G, Karpettas N, Parati G, Asmar R, Imai Y, Wang J, Mengden T, Shennan A; on behalf of the Working Group on Blood Pressure Monitoring of the European Society of Hypertension. European Society of Hypertension International Protocol revision 2010 for the Validation of Blood Pressure Measuring Devices In Adults. Blood Press Monit 2010;15:23–38 [32]
	O'Brien E, Pickering T, Asmar R, Myers M, Parati G, Staessen J, Mengden T, Imai Y, Waeber B, Palatini P with the statistical assistance of Atkins N and Gerin W on behalf of the Working Group on Blood Pressure Monitoring of the European Society of Hypertension. International protocol for validation of blood pressure measuring devices in adults. Blood Press Monit 2002;7:3-17 [33]
British Hypertension Society Protocol	O'Brien E, Petrie J, Littler WA, de Swiet M, Padfield PL, Altman D, Bland M, Coats A, Atkins N. The British Hypertension Society Protocol for the evaluation of blood pressure measuring devices. J Hypertens 1993;11(suppl 2):S43–S63 [30]
	O'Brien E, Petrie J, Littler WA, de Swiet M, Padfield PL, Altman D, Bland M, Coats A, Atkins N. Short report. An outline of the British Hypertension Society Protocol for the evaluation of blood pressure measuring devices. J Hypertens 1993;11:677–679 [31]
	O'Brien E, Petrie J, Littler W, De Swiet M, Padfield P, O'Malley K, Jamieson MJ, Altman D, Bland M, Atkins N. The Britisih Hypertension Society protocol for the evaluation of automated and semi-automated blood pressure measuring devices

(continued)

Table 4 (continued)

Standard/validation protocol	References
	with special reference to ambulatory systems. J Ambulatory Monitoring 1991;4:207–228 [28] O'Brien E, Petrie J, Littler W, de Swiet M, Padfield PL, O'Malley K, Jamieson M, Altman D, Bland M, Atkins N. The British Hypertension Society Protocol for the evaluation of automated and semi-automated blood pressure measuring devices with special reference to ambulatory systems. In Blood Pressure Measurement. Eds. E. O'Brien and K. O'Malley. Handbook of Hypertension. Eds. W.H. Birkenhager and J.L.Reid. Elsevier. Amsterdam. 1991. pp. 430-451 [29] O'Brien E, Petrie J, Littler W, de Swiet M, Padfield PL, O'Malley K, Jamieson M, Altman D, Bland M, Atkins N. The British Hypertension Society Protocol for the evaluation of automated and semi-automated blood pressure measuring devices with special reference to ambulatory systems. J Hypertens 1990;8:607–619 [27]
Association for the Advancement of Medical Instrumentation (USA)	Association for the Advancement of Medical Instrumentation. American National Standard. ANSI/AAMI/ISO 81060-2:2013 Non-invasive sphygmomanometers—Part 2: Clinical investigation of automated measurement type. 4301 N. Fairfax Drive, Suite 301, Arlington, VA 22203-1633, USA: AAMI; 2013 [26] Association for the Advancement of Medical Instrumentation. American National Standard. Manual, electronic or automated sphygmomanometers ANSI/AAMI SP10-2002/A1. 3330 Washington Boulevard, Suite 400, Arlington, VA 22201-4598, USA: AAMI; 2003 [25] Association for the Advancement of Medical Instrumentation. American National Standard. Manual, electronic or automated sphygmomanometers ANSI/AAMI SP10-2002. 3330 Washington Boulevard, Suite 400, Arlington, VA 22201-4598, USA: AAMI; 2003 [24] Association for the Advancement of Medical Instrumentation. American National Standard. Electronic or automated sphygmomanometers ANSI/AAMI SP10-1992/A1. 3330 Washington Boulevard, Suite 400, Arlington, VA 22201-4598, USA: AAMI; 1996 [23] Association for the Advancement of Medical Instrumentation. American National Standard. Electronic or automated sphygmomanometers ANSI/AAMI SP10-1992. 3330 Washington

(continued)

Table 4 (continued)

Standard/validation protocol	References
	Boulevard, Suite 400, Arlington, VA 22201-4598, USA: AAMI; 1993 [22] Association for the Advancement of Medical Instrumentation. American National Standard. Electronic or automated sphygmomanometers ANSI/AAMI SP10-1987. 3330 Washington Boulevard, Suite 400, Arlington, VA 22201-4598, USA: AAMI; 1987 [21]
Other Protocols	Normenausschuß Feinmechanik und Optik (NaFuO) im DIN Deutsches Institut für Normierung e V: Non-invasive sphygmomanometers—Clinical Investigation. Berlin: Beuth Verlag; 1996
International Organization of Legal Metrology (OIML) recommendations	OIML Recommendation R 16 Part 1 Non-invasive mechanical sphygmomanometers; 2002 [38] OIML Recommendation R 16 Part 2 Non-invasive automated sphygmomanometers; 2002 [39]

Table 5 Comparison of ANSI/AAMI SP10, BHS and DIN 58130 protocol for the evaluation of noninvasive blood pressure monitors if manual auscultatory method is used as reference

Evaluation method	ANSI/AAMI SP10 (1992)	BHS (1993)	DIN 58130 (1995)
Reference method	Manual auscultatory	Manual auscultatory	Manual auscultatory
Subjects			
Number	≥ 85	85	≥ 85
Sex	Not applicable	Not applicable	Not applicable
Age	Elderly included	Not applicable	Not applicable
Systolic pressure	<100 mmHg, >10% of subjects >180 mmHg, >10% of subjects	<90 mmHg, 8 subjects 90–129 mmHg, 20 subjects 130–160 mmHg, 20 subjects 161–180 mmHg, 20 subjects >180 mmHg, 8 subjects	80– 180 mmHg
Diastolic pressure	<60 mmHg, >10% of subjects >100 mmHg, >10% of subjects	<60 mmHg, 8 subjects 60–79 mmHg, 20 subjects 80–100 mmHg, 20 subjects 101–110 mmHg, 20 subjects >110 mmHg, 8 subjects	60– 110 mmHg

(continued)

Table 5 (continued)

Evaluation method	ANSI/AAMI SP10 (1992)	BHS (1993)	DIN 58130 (1995)
Circumference of the upper arm	<25 cm, 10% of subjects >35 cm, 10% of subjects		<24 cm or >32 cm, 25% of subjects
Number of observers	2	3	2
Stethoscope with double pairs of earpieces	Yes	Yes	Optional
Comparative measurements (no. of meas. per subject)			
Simultaneous, single arm	Preferably (≥ 3)	Preferably (3)	(≥ 3)
Simultaneous, dual arm	3. option	less suitable	($\geq 3 + 2 \times 3$)
Sequential, single arm	2. option	2. option	Less suitable
Sequential, dual-arm	Less suitable	Less suitable	Less suitable
Maximal allowable error (for systolic and diastolic pressure treated separately)	The mean ($\overline{\Delta}$) and standard deviation (s_{Δ}) of the difference between the monitor's and reference measurements must satisfy the following limits: -5 mmHg $\leq \overline{\Delta}$ ≤ 5 mmHg $s_{\Delta} \leq 8$ mmHg	The percentage of monitor's measurements that differ from the corresponding reference measurements as follows: 50% of meas., ≤ 5 mmHg 75% of meas., ≤ 10 mmHg 90% of meas., ≤ 15 mmHg	Same as for the ANSI/AAMI SP10

with a standard deviation not greater than 8 mmHg. This applies to the results from a clinical trial protocol when evaluating the accuracy of the blood pressure algorithm in a population of subjects and is not part of regular inspection.

It is recommended that only those devices that have passed validation tests of ANSI/AAMI, BHS or ESH should be used in practice. However, the fact that a device passed a validation test does not mean that it will provide accurate readings in all patients [4].

Rigorous methods such as those of the ANSI/AAMI, BHS and other standards are time consuming and impractical in the development stage of a new device or routine performance testing and may or may not be effective. A useful resource of results of validation studies which were published in peer-reviewed journals and other general and specialized journals is available at the website http://www. dableducational.org/ launched by the Working Group of the European Society of

Table 6 Comparison of the British Hypertension Society Protocol BHS (1993), the International Protocol of the European Society of Hypertension ESH IP (2002) and revised protocol ESH IP (2010)

Evaluation method	BHS (1993)	ESH IP (2002)	ESH IP (2010)
Reference method	Manual auscultatory	Manual auscultatory	Manual auscultatory
Subjects			
Number	85	33 (Phase 1: 15, Phase 2: 18)	33
Sex	Not applicable	≥ 5 M and ≥ 5 F (Phase 1) ≥ 10 M and ≥ 10 F (Phase 2)	≥ 10 M and ≥ 10 F
Age	Not applicable	≥ 30 years	≥ 25 years
Systolic pressure	<90 mmHg, 8 subjects 90–129 mmHg, 20 subjects 130–160 mmHg, 20 subjects 161–180 mmHg, 20 subjects >180 mmHg, 8 subjects	Phase 1 90–129 mmHg, 5 subjects 130–160 mmHg, 5 subjects 161–180 mmHg, 5 subjects Phase 2 90–129 mmHg, 11 subjects 130–160 mmHg, 11 subjects 161–180 mmHg, 11 subjects	90*–129 mmHg, 10–12 subjects 130–160 mmHg, 10–12 subjects 161–180* mmHg, 10–12 subjects Up to 4 recruitment pressures are permitted to be outside these limits
Diastolic pressure	<60 mmHg, 8 subjects 60–79 mmHg, 20 subjects 80–100 mmHg, 20 subjects 101–110 mmHg, 20 subjects >110 mmHg, 8 subjects	Phase 1 40–79 mmHg, 5 subjects 80–100 mmHg, 5 subjects 101–130 mmHg, 5 subjects Phase 2 40–79 mmHg, 11 subjects 80–100 mmHg, 11subjects 101–130 mmHg, 11 subjects	40*–79 mmHg, 10–12 subjects 80–100 mmHg, 10–12 subjects 101–130* mmHg, 10–12 subjects Up to 4 recruitment pressures are permitted to be outside these limits

(continued)

Table 6 (continued)

Evaluation method	BHS (1993)	ESH IP (2002)	ESH IP (2010)
Number of observers	3	2 + 1 supervisor	2 + 1 supervisor
Comparative measurements	3	3	3
Maximal allowable error (for systolic and diastolic pressure treated separately)	the percentage of monitor's measurements that differ from the corresponding reference measurements as follows: 50% of meas., ≤ 5 mmHg 75% of meas., ≤ 10 mmHg 90% of meas., ≤ 15 mmHg	Phase 1–45 meas. on 15 subjects (min requirements): At least one of 25 meas., ≤ 5 mmHg 35 meas., ≤ 10 mmHg 40 meas., ≤ 15 mmHg Phase 2.1–99 meas. on 33 subjects (min requirements): Two of 65 meas., ≤ 5 mmHg 80 meas., ≤ 10 mmHg 95 meas., ≤ 15 mmHg All of 60 meas., ≤ 5 mmHg 75 meas., ≤ 10 mmHg 90 meas., ≤ 15 mmHg Phase 2.2 At least 22 of the 33 subjects must have at least two of their three comparisons lying within 5 mmHg. At most, three of the 33 subjects can have all three of their comparisons over 5 mmHg apart	Part 1 (min requirements): Two of 73 meas., ≤ 5 mmHg 87 meas., ≤ 10 mmHg 96 meas., ≤ 15 mmHg All of 65 meas., ≤ 5 mmHg 81 meas., ≤ 10 mmHg 93 meas., ≤ 15 mmHg Part 2 At least 24 of the 33 subjects must have at least two of their three comparisons lying within 5 mmHg. At most, three of the 33 subjects can have all three of their comparisons over 5 mmHg

*Up to 4 recruitment pressures are permitted to be outside these limits

hypertension in 2004 [40]. An alternative method of evaluation is the use of a noninvasive blood pressure simulator or the so-called "surrogate arm" (arm phantom) [41, 42]. The cuff of a monitor under test is wrapped around the surrogate arm that simulates the physiological properties and characteristics of blood flow and blood pressure in the upper arm, in a way that artificial Korotkoff vibrations and oscillometric pulses are generated in an artificial artery (surrounded by different materials that simulate bone, soft tissues and skin) in response to artery occlusion. That provides simpler and more controllable testing environment, but the key issue is whether the simulator is capable to precisely simulate in vivo condition [43].

6 Inspection

Blood pressure measuring equipment should be regularly checked and calibrated. Frequency of inspections and calibrations should meet legislative and regulatory requirements and manufacturer's recommendations. Maintenance recommendations vary depending on the type, frequency and location of use (i.e. hand-held devices are likely to receive more shocks and drops than the wall or desk mounted). Faulty cuffs, hoses, aneroid gauges and mercury manometers can all lead to erroneous blood pressure measurements, with significant effects on patient care [15, 20].

Here we briefly present International Recommendations published by International Organization of Legal Metrology OIML R-16 for Non-invasive mechanical sphygmomanometers [38] and Non-invasive automated sphygmo-manometers [39] which have been issued in 2002 as separate publications. OIML International Recomendations are model regulations that establish the metrological characteristics required of certain measuring instruments and which specify methods and equipment for checking their conformity. The OIML Member States shall implement these Recommendations to the greatest possible extent.

6.1 Non-invasive Mechanical Sphygmomanometers

OIML R-16-1 specifies general, performance, efficiency and mechanical and electrical safety requirements, including test methods for type approval, for non-invasive mechanical sphygmomanometers and their accessories which, by means of an inflatable cuff, are used for the non-invasive measurement of arterial blood pressure. The application of the cuff is not limited to a particular extremity of the human body (e.g. the upper arm).

Included within the scope of this Recommendation are sphygmomanometers with a mechanical pressure sensing element and display, used in conjunction with a stethoscope or other manual methods for detecting Korotkoff sounds and for cuff inflation. Components of these devices are manometer, cuff, valve for deflation (often in combination with rapid exhaust valve), hand pump or electromechanical pump and connection hoses. These devices may also contain electro-mechanical components for pressure control.

Units of measurement

The blood pressure shall be indicated either in kilopascals (kPa) or in millimeters of mercury (mmHg).

Metrological requirements
Maximum permissible errors of the cuff pressure indication
For any set of conditions within the ambient temperature range of 15 °C to 25 °C and the relative humidity range of 20–85%, both for increasing and for decreasing pressure, the maximum permissible error for the measurement of the cuff pressure at any point of the scale range shall be ±0.4 kPa (±3 mmHg) in case of verifying the first time and ±0.5 kPa (±4 mmHg) for sphygmomanometers in use.

OIML R-16-1 also specifies requirements under storage conditions and under varying temperature conditions.

Technical requirements

Technical requirements for the cuff and bladder
The cuff shall contain a bladder. For reusable cuffs the manufacturer shall indicate the method for cleaning in the accompanying documents.

The optimum bladder size is one with dimensions such that its width is 40% of the limb circumference at the midpoint of the cuff application, and its length is at least 80%, preferably 100%, of the limb circumference at the midpoint of cuff application. Use of the wrong size can affect the accuracy of the measurement.

Technical requirements for the pneumatic system
Air leakage

Air leakage shall not exceed a pressure drop of 0.5 kPa/min (4 mmHg/min).
Pressure reduction rate

Manually operated deflation valves shall be capable of adjustment to a deflation rate from 0.3 to 0.4 kPa/s (2–3 mmHg/s). Manually operated deflation valves shall be easily adjusted to these values.
Rapid exhaust

During the rapid exhaust of the pneumatic system, with the valve fully opened, the time for the pressure reduction from 35 to 2 kPa (260–15 mmHg) shall not exceed 10 s.

Technical requirements for the pressure indicating devices
Nominal range and measuring range

The nominal range shall be equal to the measuring range. The nominal range for the cuff gauge pressure shall extend from 0 kPa to at least 35 kPa (0 mmHg to at least 260 mmHg).
Analogue indication

The scale shall be designed and arranged so that the measuring values can be read clearly and are easily recognized. The graduation shall begin with the first scale mark at 0 kPa (0 mmHg). The scale interval shall be: 0.2 kPa for a scale graduated in kPa; or 2 mmHg for a scale graduated in mmHg. Each fifth scale mark shall be indicated by greater length and each tenth scale mark shall be numbered. The distance between adjacent scale marks shall be not less than 1.0 mm. The thickness of the scale marks shall not exceed 20% of the smallest scale spacing. All scale marks shall be of equal thickness.

Additional technical requirements for mercury manometers

Internal diameter of the tube containing mercury

The nominal internal diameter of the mercury tube shall be at least 3.5 mm. The tolerance on diameter shall not exceed ±0.2 mm.

Portable devices

A portable device shall be provided with an adjusting or locking mechanism to secure it in the specified position of use.

Devices to prevent mercury from being spilled during use and transport

A device shall be placed in the tube to prevent mercury from being spilled during use and transport (for example: stopping device, locking device, etc.). This device shall be such that when the pressure in the system drops rapidly from 27 to 0 kPa (from 200 to 0 mmHg), the time taken for the mercury column to fall from 27 to 5 kPa (from 200 to 40 mmHg) shall not exceed 1.5 s. This time is known as the "exhaust time".

Quality of the mercury

The mercury shall have a purity of not less than 99.99% according to the declaration of the supplier of the mercury. The mercury shall exhibit a clean meniscus and shall not contain air bubbles.

Graduation of the mercury tube

Graduations shall be permanently marked on the tube containing mercury. If numbered at each fifth scale mark, the numbering shall be alternately on the right- and left-hand side of, and adjacent to, the tube.

Additional technical requirements for aneroid manometers

Scale mark at zero

If a tolerance zone is shown at zero it shall not exceed ±0.4 kPa (±3 mmHg) and shall be clearly marked. A scale mark at zero shall be indicated. Note: Graduations within the tolerance zone are optional.

Zero

The movement of the elastic sensing element including the pointer shall not be obstructed within 0.8 kPa (6 mmHg) below zero. Neither the dial nor the pointer shall be adjustable by the user.

Pointer

The pointer shall cover between 1/3 and 2/3 of the length of the shortest scale mark of the scale. At the place of indication it shall be not thicker than the scale mark. The distance between the pointer and the dial shall not exceed 2 mm.

Hysteresis error

The hysteresis error throughout the pressure range shall be within the range 0–0.5 kPa (0–4 mmHg).

Construction and materials

The construction of the aneroid manometer and the material for the elastic sensing elements shall ensure an adequate stability of the measurement. The elastic sensing elements shall be aged with respect to pressure and temperature. After 10,000 alternating pressure cycles the change in the pressure indication of the aneroid manometer shall be not more than 0.4 kPa (3 mmHg) throughout the pressure range.

Safety requirements
Resistance to vibration and shock
The sphygmomanometer shall comply with the relevant paragraphs of International Document OIML D 11.

After testing, the device shall comply with the requirements of Maximum permissible errors of the cuff pressure indication (of this Recommendation).
Mechanical safety
It shall be possible to abort the blood pressure measurement at any time by activating the manual rapid exhaust valve, which shall be easily accessible.
Tamper proofing
Tamper proofing of the manometer shall be achieved by requiring the use of a tool or breaking a seal.
Electrical safety
Regional or national regulations may specify electrical safety requirements.

For each requirement test procedures are also described (see Annex A of OIML R 16-1). Test report format is also given (seen Annex B of OIML R 16-1).

Metrological controls
Requirements related to type approval, verification (initial and subsequent), sealing, marking the device and manufacturer's information are also prescribed.

6.2 Non-invasive Automated Sphygmomanometers

OIML R-16-2 specifies general, performance, efficiency and mechanical and electrical safety requirements, including test methods for type approval, for non-invasive electronic or automated sphygmomanometers and their accessories which, by means of an inflatable cuff, are used for the non-invasive measurement of arterial blood pressure. This Recommendation only applies to devices measuring at the upper arm, the wrist or the thigh.

Units of measurement
The blood pressure shall be indicated either in kilopascals (kPa) or in millimeters of mercury (mmHg).

Metrological requirements

Maximum permissible errors of the cuff pressure indication For any set of conditions within the ambient temperature range of 15–25 °C and the relative humidity range of 20–85%, both for increasing and for decreasing pressure, the maximum permissible error for the measurement of the cuff pressure at any point of the scale range shall be ±0.4 kPa (±3 mmHg) in case of verifying the first time and ±0.5 kPa (±4 mmHg) for sphygmomanometers in use.

Maximum permissible errors of the overall system as measured by clinical tests (this is carried out by the manufacturer).

The following maximum permissible errors shall apply for the overall system: maximum mean error of measurement: ±0.7 kPa (±5 mmHg); maximum experimental standard deviation: 1.1 kPa (8 mmHg). For recommended test methods see validation protocols of ANSI/AAMI, BHS, ESH described in Sect. 5.5.

Environmental performance

Environmental conditions under which blood pressure measuring system shall maintain the specified requirements are detailed. See appropriate section in the Recommendation [39].

Technical requirements

General

Equipment, or parts thereof, using materials or having forms of construction different from those detailed in this Recommendation shall be accepted if it can be demonstrated that an equivalent degree of safety and performance is obtained.

Technical requirements for the cuff and bladder

The cuff shall contain a bladder. For reusable cuffs the manufacturer shall indicate the method for cleaning in the accompanying documents. Note: The optimum bladder size is one with dimensions such that its width is 40% of the limb circumference at the midpoint of the cuff application and its length is at least 80%, preferably 100% of the limb circumference at the midpoint of cuff application. Use of the wrong size can affect the accuracy of the measurement.

Technical requirements for the display

The display shall be designed and arranged so that the information including measuring values can be read and easily recognized. Testing shall be carried out by visual inspection.

If abbreviations are used on the display they shall be as follows: S or SYS: systolic blood pressure (value); D or DIA: diastolic blood pressure (value); M or MAP: mean arterial blood pressure (value). Single letter abbreviations shall be positioned in such a way to avoid confusion with SI units.

Effect of voltage variations of the power source

Internal electrical power source

Changes of the voltage within the working range determined according to Test methods for the effect of voltage variations of the power source on the cuff pressure indication (as described in Annex A of this Recommendation) shall not influence the cuff pressure reading and the result of the blood pressure measurement.

Outside this working range no cuff pressure reading and no result of the blood pressure measurement shall be displayed.

External electrical power source

Changes of the voltage within the working range specified by the manufacturer shall not influence the cuff pressure reading and the result of the blood pressure measurement.

Incorrect values resulting from voltage variations outside these limits shall not be displayed.

Note: In the case of any malfunction of the equipment, deflation to below 2 kPa (15 mmHg) must be guaranteed within 180 s in the case of adult patients and to below 0.7 kPa (5 mmHg) within 90 s in the case of neonatal/infant patients.

Pneumatic system

Air leakage

Air leakage shall not exceed a pressure drop of 0.8 kPa/min (6 mmHg/min).

Pressure reducing system for devices using the auscultatory method

The pressure reducing system for manually operated and automated deflation valves shall be capable of maintaining a deflation rate of 0.3–0.4 kPa/s (2–3 mmHg/s) within the target range of systolic and diastolic blood pressure. For devices which control the pressure reduction as a function of the pulse rate, a deflation rate of 0.3 kPa/pulse to 0.4 kPa/pulse (2–3 mmHg/pulse) shall be maintained.

Rapid exhaust

During the rapid exhaust of the pneumatic system, with the valve fully opened, the time for the pressure reduction from 35 to 2 kPa (260–15 mmHg) shall not exceed 10 s. For blood pressure measuring systems having the capability to measure in a neonatal/infant mode, the time for the pressure reduction from 20 to 0.7 kPa (150–5 mmHg) during the rapid exhaust of the pneumatic system with the valve fully opened shall not exceed 5 s.

Zero setting

Blood pressure measuring systems shall be capable of automatic zero setting. The zero setting shall be carried out at appropriate intervals, at least starting after switching on the device. At the moment of the zero setting a gauge pressure of 0 kPa (0 mmHg) shall exist and be displayed thereafter. Devices performing zero setting only immediately after switching on, shall switch off automatically when the drift of the pressure transducer and the analog signal processing exceeds 0.1 kPa (1 mmHg).

Electromagnetic compatibility

Either: electrical and/or electromagnetic interferences shall not lead to degradations in the cuff pressure indication or in the result of the blood pressure measurement; or if electrical and/or electromagnetic interferences lead to an abnormality, the abnormality shall be clearly indicated and it shall be possible to restore normal operation within 30 s after cessation of the electromagnetic disturbance. Testing should be carried out in accordance with the relevant OIML provisions.

Stability of the cuff pressure indication

The change in the cuff pressure indication shall not be more than 0.4 kPa (3 mmHg) throughout the pressure range after 10,000 simulated measurement cycles.

Pressure indicating device

Nominal range and measuring range

The nominal range for the cuff pressure measurement shall be specified by the manufacturer. The measuring and indication ranges of the cuff pressure shall be equal to the nominal range. Values of blood pressure measurement results outside the nominal range of cuff pressure shall be clearly indicated as out of range.

Digital indication

The digital scale interval shall be 0.1 kPa (1 mmHg). If the measured value of a parameter is to be indicated on more than one display, all the displays shall indicate the same numerical value. Measured numerical values on the display(s), and the symbols defining the units of measurement shall be arranged in such a way so as to avoid misinterpretation. Numbers and characters should be clearly legible.

Signal input and output ports

The construction of the signal input and output ports (excluding internal interfaces, e.g. microphone signal input) relevant to the non-invasive blood pressure measurement shall ensure that incorrectly fitted or defective accessories shall not result in erroneous indication of cuff pressure or erroneous indication of blood pressure.

Alarms

If alarms are used they shall be of at least medium priority.

Safety

Cuff pressure

It shall be possible to abort any blood pressure measurement at any time by single key operation and this shall lead to a rapid exhaust.

Unauthorized access

All controls which affect accuracy shall be sealed against unauthorized access.

Tubing connectors

Users of equipment intended for use in environments employing intervascular fluid systems shall take all necessary precautions to avoid connecting the output of the blood pressure measuring device to such systems as air might inadvertently be pumped into a blood vessel if, for example, Luer locks were used.

Electrical safety

Electronic or automated sphygmomanometers shall comply with the relevant national safety regulations.

Resistance to vibration and shock

The sphygmomanometer shall comply with the relevant provisions of OIML D 11.

After testing, the device shall comply with the requirements of Maximum permissible errors of the cuff pressure indication (of this Recommendation).

Metrological controls

Requirements related to type approval, verification (initial and subsequent), sealing, marking the device and manufacturer's information are also prescribed.

For each requirement test procedures are also described (see Annex A of OIML R 16-2). Test report format is also given (see Annex B of OIML R 16-2).

In order to speed up the testing procedure in everyday practice blood-pressure simulations are available (e.g. BP Pump 2 NIBP Blood Pressure Simulator by Fluke) as well as Electrical Safety Analyzers to test for IEC60601-1 compliance.

7 Summary

To be suitable for clinical use blood pressure measuring devices must comply with numerous requirements that depend on state legislative. To ensure reliable and accurate blood pressure measurement it is equally important that in hospitals and other medical facilities quality assurance (QA) measures have been implemented. It is also necessary to perform routine inspection and calibration of blood pressure manometers, and all automated blood pressure measuring devices used in hospitals and primary care facilities. How often, by whom, and at what cost remain to be decided by responsible authorities. Training of those who use blood pressure measuring devices must be done and kept up to date.

Devices for home use are rarely thoroughly tested. Especially automatic oscillometric blood pressure monitors that are widely available could in some patients show unreliable and highly inaccurate results. It is therefore important to inform patients of the limitation of these devices. These limitations are also present in oscillometric monitors used in clinics, but physicians and nurses could cope with them if properly trained.

References

1. Levick JR (2010) An introduction to cardiovascular physiology, 5th edn. Hodder Arnold, London
2. Chobanian AV, Bakris GL, Black HR, Cushman WC, Green LA, Izzo Jr. JL, Jones DW, Materson BJ, Oparil S, Wright Jr. JT, Roccella EJ et al (2003) Seventh report of the Joint National Committee on Prevention, Detection, Evaluation, and Treatment of High Blood Pressure. Hypertension. Joint National Committee On Prevention 42(6):1206–1252. doi:10. 1161/01.HYP.0000107251.49515.c2
3. Raised blood pressure. World Health Organization. Global Health Observatory (GHO) data. http://www.who.int/gho/ncd/risk_factors/blood_pressure_text/en/
4. Pickering TG, Hall JE, Appel LJ, Falkner BE, Graves J, Hill MN, Jones DW, Kurtz T, Sheps SG, Roccella EJ (2005) Recommendations for blood pressure measurement in humans and experimental animals: part 1: blood pressure measurement in humans: a statement for professionals from the subcommittee of professional and public education of the American heart association council on high blood pressure research. Circulation 111(5):697–716
5. Geddes LA (1991) Handbook of blood pressure measurement. Humana Press, Clifton
6. Ng KG, Small CF (1994) Survey of automated noninvasive blood pressure monitors. J Clin Eng 19(6):452–475
7. Drzewiecki GM, Melbin J, Noordergraaf A (1989) The Korotkoff sound. Ann Biomed Eng 17 (4):325–359

8. London SB, London RE (1967) Comparison of indirect pressure measurement (Korotkoff) with simultaneous direct brachial artery pressure distal to the cuff. Adv Inter Med 13:127–142

9. Breit SN, O'Rourke MF (1974) Comparison of direct and indirect arterial pressure measurements in hospitalized patients. Aust N Z J Med 4(5):485–491

10. Geddes LA, Voelz M, Combs C, Reiner D, Babbs CF (1982) Characterization of the oscillometric method for measuring indirect blood pressure. Ann Biomed Eng 10(6):271–280

11. Drzewiecki G, Hood R, Apple H (1994) Theory of the oscillometric maximum and the systolic and diastolic detection ratios. Ann Biomed Eng 22(1):88–96

12. Geddes LA, Voelz M, James S, Reiner D (1981) Pulse arrival time as a method of obtaining systolic and diastolic blood pressure indirectly. Med Biol Eng Comput 19(5):671–672

13. Yamakshi K, Kamiya A, Shimazu H, Ito H, Togawa T (1983) Noninvasive automatic monitoring of instantaneous arterial blood pressure using the vascular unloading technique. Med Biol Eng Comput 21(5):557–565

14. Sato T, Nishinaga M, Kawamoto A, Ozawa T, Takatsuji H (1993) Accuracy of a continuous blood pressure monitor based on arterial tonometry. Hypertension 21(6):866–874

15. Blood pressure measurement devices v2.1 Medicines and Healthcare Products Regulatory Agency (MHRA) (2013) https://www.gov.uk/government/publications/blood-pressure-measurement-devices

16. Mion D, Pierin AM (1998) How accurate are sphygmomanometers? J Hum Hypertens 12:245–248

17. Markandu ND, Whitcher F, Arnold A, Carney C (2000) The mercury sphygmomanometer should be abandoned before it is proscribed. J Hum Hypertens 14:31–36

18. Yarows SA, Qian K (2001) Accuracy of aneroid sphygmomanometers in clinical usage: University of Michigan experience. Blood Press Monit 6:101–106

19. Canzanello VJ, Jensen PL, Schwartz GL (2001) Are aneroid sphygmomanometers accurate in hospital and clinic settings? Arch Intern Med 161:729–731

20. Coleman AJ, Steel SD, Ashworth M, Vowler SL, Shennan A (2005) Accuracy of the pressure scale of sphygmomanometers in clinical use within primary care. Blood Press Monit 10 (04):181–188

21. Association for the Advancement of Medical Instrumentation (1987) American National Standard. Electronic or automated sphygmomanometers ANSI/AAMI SP10-1987. 3330 Washington Boulevard, Suite 400, Arlington, VA 22201-4598, USA: AAMI

22. Association for the Advancement of Medical Instrumentation (1993) American National Standard. Electronic or automated sphygmomanometers ANSI/AAMI SP10-1992. 3330 Washington Boulevard, Suite 400, Arlington, VA 22201-4598, USA: AAMI

23. Association for the Advancement of Medical Instrumentation (1996) American National Standard. Electronic or automated sphygmomanometers ANSI/AAMI SP10-1992/A1. 3330 Washington Boulevard, Suite 400, Arlington, VA 22201-4598, USA: AAMI

24. Association for the Advancement of Medical Instrumentation (2003) American National Standard. Manual, electronic or automated sphygmomanometers ANSI/AAMI SP10-2002. 3330 Washington Boulevard, Suite 400, Arlington, VA 22201-4598, USA: AAMI

25. Association for the Advancement of Medical Instrumentation (2003) American National Standard. Manual, electronic or automated sphygmomanometers ANSI/AAMI SP10-2002/A1. 3330 Washington Boulevard, Suite 400, Arlington, VA 22201-4598, USA: AAMI

26. Association for the Advancement of Medical Instrumentation (2013) American National Standard. ANSI/AAMI/ISO 81060-2:2013 Non-invasive sphygmomanometers—part 2: clinical investigation of automated measurement type. 4301 N. Fairfax Drive, Suite 301, Arlington, VA 22203-1633, USA: AAMI

27. O'Brien E, Petrie J, Littler W, de Swiet M, Padfield PL, O'Malley K, Jamieson M, Altman D, Bland M, Atkins N (1990) The British Hypertension Society Protocol for the evaluation of automated and semi-automated blood pressure measuring devices with special reference to ambulatory systems. J Hypertens 8:607–619

28. O'Brien E, Petrie J, Littler W, De Swiet M, Padfield P, O'Malley K, Jamieson MJ, Altman D, Bland M, Atkins N (1991) The British Hypertension Society protocol for the evaluation of automated and semi-automated blood pressure measuring devices with special reference to ambulatory systems. J Ambul Monit 4:207–228

29. O'Brien E, Petrie J, Littler W, de Swiet M, Padfield PL, O'Malley K, Jamieson M, Altman D, Bland M, Atkins N (1991) The British Hypertension Society Protocol for the evaluation of automated and semi-automated blood pressure measuring devices with special reference to ambulatory systems. In: O'Brien E, O'Malley K (eds) Blood pressure measurement. Birkenhager WH, Reid JL (eds) Handbook of hypertension. Elsevier, Amsterdam, pp 430–451

30. O'Brien E, Petrie J, Littler WA, de Swiet M, Padfield PL, Altman D, Bland M, Coats A, Atkins N (1993) The British Hypertension Society Protocol for the evaluation of blood pressure measuring devices. J Hypertens 11(suppl 2):S43–S63

31. O'Brien E, Petrie J, Littler WA, de Swiet M, Padfield PL, Altman D, Bland M, Coats A, Atkins N (1993) Short report. An outline of the British Hypertension Society Protocol for the evaluation of blood pressure measuring devices. J Hypertens 11:677–679

32. O'Brien E, Atkins N, Stergiou G, Karpettas N, Parati G, Asmar R, Imai Y, Wang J, Mengden T, Shennan A; on behalf of the Working Group on Blood Pressure Monitoring of the European Society of Hypertension (2010) European Society of Hypertension International Protocol revision 2010 for the validation of blood pressure measuring devices in adults. Blood Press Monit 15:23–38

33. O'Brien E, Pickering T, Asmar R, Myers M, Parati G, Staessen J, Mengden T, Imai Y, Waeber B (2002) Palatini P with the statistical assistance of Atkins N and Gerin W on behalf of the Working Group on Blood Pressure Monitoring of the European Society of Hypertension. International protocol for validation of blood pressure measuring devices in adults. Blood Press Monit 7:3–17

34. DIN 58130 Nichtinvasive Blutdruckmeßgeräte - Klinische Prüfung (1995)

35. European Standard EN 1060-1:1996. Specification for Non-invasive sphygmomanometers. Part I. General requirements. 1995. European Commission for Standardisation. Rue de Stassart 36, B-1050 Brussels

36. European Standard EN 1060-2:1996. Specification for Non-invasive sphygmomanometers. Part 2. Supplementary requirements for mechanical sphygmomanometers. 1995. European Commission for Standardisation. Rue de Stassart 36, B-1050 Brussels

37. European Standard EN 1060-3: 1997. Non-invasive sphygmomanometers. Part 3. Supplementary requirements for electro-mechanical blood pressure measuring systems. European Committee for Standardization. Rue de Stassart 36, B-1050, Brussels

38. International Organization of Legal Metrology. OIML Recommendation R 16 Part 1 Non-invasive mechanical sphygmomanometers (2002) http://www.oiml.org/en/files/pdf_r/r016-p-e02.pdf

39. International Organization of Legal Metrology. OIML Recommendation R 16 Part 2 Non-invasive automated sphygmomanometers (2002) http://www.oiml.org/en/files/pdf_r/r016-p-e02.pdf

40. http://www.dableducational.org/

41. Ng KG, Small CF (1992) Review of methods & simulators for evaluation of noninvasive blood pressure monitors. J Clin Eng 17(6):469–479

42. Yong P, Geddes LA (1990) A surrogate arm for evaluating the accuracy of instruments for indirect measurement of blood pressure. Biomed Instrum Technol 24(2):130–135

43. Amoore J, Vacher E, Murray I (2006) Can a simulator that regenerates physiological waveforms evaluate oscillometric non-invasive blood pressure devices? Blood Press Monit 11:63–67

Inspection and Testing of Diagnostic Ultrasound Devices

Gordana Žauhar, Ana Diklić and Slaven Jurković

Abstract This chapter provides an overview of the safety aspects of application of ultrasound in medicine. It starts with the short history of ultrasound methods and devices as well as basic principles of ultrasound imaging systems. The application of ultrasound in medicine greatly evolved and nowadays it can be divided into two main areas: imaging and therapy. In order to assure a safe and responsible application of ultrasound in medicine one should be aware of physical processes which can be produced in tissue by ultrasound such as temperature rise, cavitation and acoustic streaming. The importance of understanding how these processes can affect the human cell is self-explanatory. In order to better understand the guidelines for testing and quality control of ultrasonic devices it is necessary to give an overview of basic output parameters. Only the most important parameters from the point of safe use of ultrasound are described, e.g. acoustic pressure, acoustic power and intensity. In order to protect the public against inappropriate exposure when ultrasound is used for medical applications, international standards and national regulations are developed. Diagnostic ultrasound imaging is very often the basis for diagnostic decision; therefore it is also necessary to include such systems into a comprehensive quality assurance programme. Ultrasound systems used for therapy have larger intensities though there are additional safety requirements compared to diagnostic systems. The ultrasound intensity, effective radiation area and beam non-uniformity ratio and are parameters which should be monitored.

G. Žauhar (✉)
Medical Faculty and Department of Physics, University of Rijeka, Rijeka, Croatia
e-mail: gordana.zauhar@medri.uniri.hr

A. Diklić
Department of Medical Physics, Clinical Hospital Centre Rijeka, Rijeka, Croatia

S. Jurković
Department of Medical Physics, Clinical Hospital Centre Rijeka and Medical Faculty
University of Rijeka, Rijeka, Croatia

© Springer Nature Singapore Pte Ltd. 2018
A. Badnjević et al. (eds.), *Inspection of Medical Devices*, Series in Biomedical
Engineering, https://doi.org/10.1007/978-981-10-6650-4_6

1 History of Ultrasound Methods and Devices

The history of ultrasound and ultrasound waves can be tracked since 1790. When Lazzaro Spallanzani, an Italian biologist, discovered the ability of bats navigating accurately in the dark through echo reflection from high frequency sound.

In 1826, a Swiss physicist and engineer Jean-Daniel Colladon used a church bell under water to prove that sound travels faster through water than air. This experiment almost exactly determined the speed of sound in water which opened up great possibilities in the field of ultrasound.

A real revolution happened somewhat later in 1881. When Pierre and Jacques Curie noted that the electricity is created in the crystal of quartz under mechanical stress. This phenomenon was described as the piezoelectric effect and it led to the development of ultrasound probes as we know today.

Two events triggered the application of ultrasound for detection of objects. These are: When Titanic sank in 1912 and the beginning of World War I. During World War I, the French physicist, Paul Langevin began working on the use of ultrasound to detect submarines and icebergs through echo location. Paul Langevin was a doctoral student of Pierre Curie and he is famous for his two US patents with Constantin Chilowsky in 1916 and 1917 involving ultrasonic submarine detection [1]. However, World War I was over by the time their invention was operational.

The medical use of ultrasound actually started for therapy purposes. The destructive ability of high intensity ultrasound has been recognized already by Langevin in 1920. High intensity ultrasound has developed soon to become a tool in neurosurgery. Also, ultrasound effects were widely used in physical and rehabilitation medicine.

Karl Dussik, Austrian neurologist and psychiatrist, was the first one to use ultrasound waves to diagnose brain tumour by the end of the 1930s. He called the procedure "hyperphonography".

George Ludwig and his team of co-workers were the first to record and describe the speed difference of the sound waves going through different tissues and organs in animals, in 1947. This was a huge progress in the field of medical ultrasound diagnostics.

Scottish scientist Ian Donald invented and upgraded a lot of devices that were used for pregnancy diagnostics and pathology. During World War II he developed the technology of radars and sonars. In 1950s he became famous when he met a patient with an inoperabile abdominal tumour. By using new technology he discovered that the tumour was really an ovary cyst and the patient was successfully operated. Not long after that he became the "father" of gyneological ultrasound techniques. He also invented the B-mode.

During the 1950s and 1960s Douglass Howry and Joseph Holmes advanced the technology of 2D B-mode ultrasound. Until then, the patient had to be inside a water bath to perform an exam. The invention of ultrasound probe which is in full contact with the patient opened the path towards the development of modern ultrasound systems. John Wild and John Reid modified standard ultrasound systems and created a handheld B-mode device in order to provide different angles which was extremely important for breast imaging.

Although the application of ultrasound in the diagnosis began in the mid 1950s, rapid expansion of its use began in the early 1970s with introduction to two-dimensional real-time ultrasonic scanners. An additional step forward was the appearance of phased array systems in the early 1980s.

In 1970s the Doppler effect was used to construct a device for the visualization of blood circulation. A new milestone occurred with the introduction of color flow-imaging systems in the mid 1980s. The first 3D image was captured in 1980s by Kazunori Baba in Tokyo while 3D devices appeared in practice in the 1990s. Further improvements led to the implementation of 4D (real-time) capabilities.

By the development of electronics and computers, devices are continuously evolving and becoming more convenient. Contemporary ultrasound devices are completely different from the historical ones, but the goal remained the same—better diagnosis or therapy.

2 Basic Principles of Ultrasound Imaging Systems

Ultrasound imaging instruments have evolved over the last 50 years from relatively simple hand moved scanners to rather sophisticated imaging systems. In this chapter basic components and principles of a generalized ultrasound imaging system will be presented.

A block diagram of a typical ultrasound imaging system is shown in Fig. 1. A primary component of an ultrasound imaging systems is a transducer. Ultrasound transducers employ piezoelectric crystals for the generation of ultrasound. Piezoelectric crystals change size and shape when a voltage is applied and vice versa. AC voltage makes them oscillate at the same frequency and produce ultrasonic sound. The ideal piezoelectric material for medical ultrasound transducers should be both an efficient producer and sensitive receiver of ultrasonic waves. The piezoelectric materials that are commonly used in medical ultrasonic transducers are

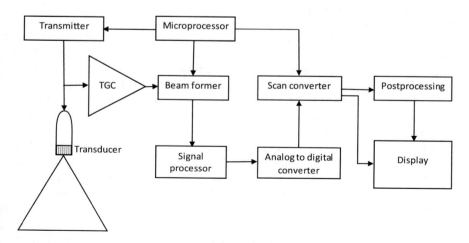

Fig. 1 Block diagram of a typical ultrasound imaging system

piezoceramics such as lead zirconate titanate (PZT) or polymer materials such as polyvinylidene fluoride (PVDF).

A transducer, which converts electrical signals to mechanical forces, generates pulses of ultrasound which are sent through a body. Organ boundaries and different types of tissues produce echoes that return back and are detected by the same transducer, which converts the acoustic signal to an electrical signal. The majority of medical ultrasound imaging systems uses the same transducer for both generation and reception of ultrasound. The ultrasound imaging system then processes the echoes and presents them usually on a grey scale image on a display. The time of return of these pulses gives information about the location of a reflector. Each point in the image corresponds to the location, and its brightness corresponds to the echo strength (hence the name for the basic ultrasonic imaging mode, B-mode, with B for brightness).

In order to obtain a two dimensional image the transmitted beam must be steered (scanned or swept). The transducer can be fabricated in a number of ways to perform B-mode imaging. In its simplest form, the B-mode transducer is a circular single-element transducer with a fixed geometric focus. Nowadays, this type of a transducer has been mostly replaced by more sophisticated multi-element transducers. Early ultrasound systems used single-element transducers that were manually scanned within a body. Modern systems use either mechanical or electrical means to scan the beam. With respect to steering methods, the great majority of instruments available today are electronically steered.

The block diagram in Fig. 1 shows the signal processing steps required for B-mode image formation. The actual implementations vary considerably among manufacturers and the types of systems.

The beam former is part of the ultrasound imaging system where the action starts. It consists of pulser, pulse delays, transmit/receive switch, amplifiers and a summer [2]. The pulser generates voltages that drive the transducer. The frequency of the voltage pulse determines the frequency of the resulting ultrasound which ranges from 2 to 15 MHz for most applications. In order to avoid echo misplacement, all echoes from one pulse must be received before the next pulse is emitted. For deeper imaging, echoes take longer to return, therefore the pulse repetition frequency (PRF) must be reduced. The pulser adjusts the PRF appropriately for the current imaging depth. Beam formation can be considered to be composed of two separate processes: beam steering and focusing [3]. The implementation of these two functions may or may not be separated, depending on the system design. The sequencing, phase delays, and variation in pulse amplitude that are necessary for electronic control of beam steering and focusing must be accomplished. The pulser and pulse delays carry out all these tasks. The transmit/receive (T/R) switches are used to direct the high voltages from the pulser and pulse delays to the transducer during transmission and then direct the returning echo voltages from the transducer to the amplifiers during reception stage.

The time gain compensation (TGC) equalises differences in received echo amplitudes caused by different reflector depths. TGC compensates for the effect of attenuation caused by the propagation of sound in tissue. The dynamic range available from typical TGC amplifiers is in the order of 60 dB.

After amplification the echo voltages pass through analog-to-digital converters (ADC). An ADC converts voltage from analog to digital form.

The signal processor performs filtering, demodulation and compression of echo data. Demodulation is the conversion of echo voltages from radio frequency form to video form. There is a large range of received echo amplitudes from the lowest level that can be detected to the maximum signal level. In order to display all echoes it is necessary to reduce the large range of received echoes. Compression is the process of decreasing the difference between the smallest and largest echo amplitude to a usable range.

The major function of the scan converter is to properly locate echo data into image or into video pixel space. The image processor converts echo data from digital to analog form and sends them to the display. The information delivered to the display can be presented in several ways. The most commonly used is brightness mode (B-mode), but in addition motion mode (M-mode) is sometimes used in echo cardiology, and occasionally amplitude mode (A-mode) in ophthalmology. The development of computers made it possible to obtain three-dimensional ultrasound imaging (3D). Ultrasound 3D imaging is accomplished by acquiring many parallel two-dimensional (2D) images and then post processing this 3D volume of echo information and presenting it in an appropriate way on 2D display. Further development of ultrasound technology led to the so called "4D imaging". Actually it is a real-time 3D imaging. 4D ultrasound imaging allows users to study the motion of various moving organs of the body. It is called 4D because time is considered as the fourth dimension.

3 Safety and Quality Assurance for Ultrasound Medical Devices

Ultrasound medical devices play an important role in diagnostics and treatment of patients. Medical ultrasound is a non-invasive, real-time, tomographic soft-tissue imaging modality, with a wide range of clinical applications, both as primary modality or as modality complement to other diagnostic procedures. Advancements in the field of diagnostic ultrasound have led to increased use of this modality in many clinical applications.

Ultrasound is often considered the preferred imaging modality because of its ability to provide continuous, real-time images without the risk of ionizing radiation and at significantly lower costs than computed tomography or magnetic resonance imaging. As with any modality, an increase in use is accompanied by an increased need for performance testing in order to ensure the repeatability and accuracy of the results. Since the final image is the basis for diagnostic decisions, the image quality produced by a scanner provides the most important information in testing scanner performance.

Testing and manufacturing guidelines for medical ultrasound devices are laid out in the medical devices directive MDD 93/42, as well as the recommendations for electrical medical equipment in IEC 60601-2-5 [4], IEC 60601-2-37 [5], and IEC 60601-2-62 [6]. The majority of the guidelines that are being followed are those related to the manufacturing and sales (as well as obtaining the CE mark), while the precision and accuracy of medical ultrasound devices in the subsequent period of use are neither sufficiently monitored nor obligatory as it is with ionizing radiation.

3.1 The Acoustic Output Parameters of Ultrasound Medical Devices

In order to better understand the guidelines for testing and quality control of ultrasonic devices it is necessary to give an overview of basic output parameters. Although there are a lot of physical quantities and parameters which are used to describe the acoustic field only the most important parameters from the point of safe use of ultrasound will be described. These acoustic output parameters are: the maximum negative or rarefaction acoustic pressure (p_- or p_r), the spatial-peak temporal-average intensity (I_{SPTA}), the temporal-average acoustic power (P), and the temperature of the transducer face (T_{surf}).

3.2 Acoustic Pressure

When ultrasound propagates through a medium, it induces a series of compressions and rarefactions of the particles constituting the medium and because of that acoustic pressure varies with time. Acoustic pressure is normally measured in water using hydrophone. Figure 2 shows a typical pressure waveform obtained by hydrophone. The most important parameters which can be derived directly from the pressure waveform are: maximum positive or peak compression acoustic pressure, p_+ (or p_c) and maximum negative or peak rarefaction acoustic pressure, p_- (or p_r). The greatest value of the acoustic pressure is of considerable importance in assessing the risk of a cavitation occurrence. Acoustic cavitation is the formation and collapse of gaseous and vapour bubbles in the medium due to an acoustic pressure field. In particular, the peak rarefaction pressure is strongly related to cavitation events. The peak rarefaction pressure changes with position in the beam and is greatest in the focal region.

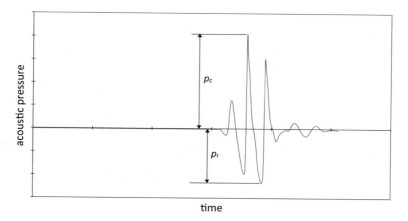

Fig. 2 Acoustical pressure parameters: maximum positive or peak compression acoustic pressure, and maximum negative or peak rarefaction acoustic pressure

3.3 Acoustic Power

When passing through the tissue ultrasound waves carry energy that is gradually absorbed and deposited in the tissue. The power of an ultrasound beam is the total energy passing through the whole cross-sectional area of the beam per unit of time. It is measured in watts (W). Acoustic power is a measure of the rate at which the transducer emits energy. A medium which absorbs the acoustic energy, of course heats up, but it is also subject to a radiation force.

In ultrasound fields at megahertz frequencies, output power is typically determined by measuring the force on a target using a radiation force balance. The principle of the radiation force measurement technique is that a target placed in the path of an acoustic beam will experience a radiation force (F). The target may be of an absorbing or reflecting type. For a plane wave incident on a perfectly absorbing target, the acoustic power (P) is given by:

$$P = F \cdot c \tag{1}$$

where F is the radiation force, c is propagation speed of ultrasound in the coupling medium, usually water. However, the relationship between the radiation force and the output power is affected by the focusing or other geometrical aspects of the field, by the type and shape of the target, by the distance of the target from the transducer, by the absorption of high frequency harmonics associated with the non-linear propagation of ultrasound waves in water, and by the acoustic streaming currents. Fortunately many of these effects are small for typical diagnostic or physiotherapy ultrasound fields and can be ignored in the first approximation.

The radiation force method has become the internationally accepted method for characterizing ultrasonic power. There are a number of variations in the design of

power measurement systems. Either the target or the transducer can be kept fixed
and the force detected at the non-fixed element. The radiation force is typically
determined by measuring the change in weight of an initially buoyant target. It can
be calculated that one watt of acoustic power produces a radiation force and the
change in weight of approximately 69 mg on an absorbing target. Therefore, it is
clear that measuring of acoustic power requires the application of a sensitive
microbalance and measurement should be performed under strictly controlled
conditions. Unfortunately, at the present time there is no commercially available
form of power measuring device that is both sufficiently sensitive and portable for
measurements at hospital sites. However, the two power balances designed by
Perkins [7] and Farmery and Whittingham [8] have been found previously to be
satisfactory and suitable for measurements in hospital environment.

3.4   Intensity

Although acoustic power is important, it is essential to describe how that power is
distributed throughout the beam and across a scanning plane. This is measured as
the acoustic intensity of ultrasound, which can be described as the power transferred
per unit area, where the area is an imagined surface that is perpendicular to the
direction of propagation of the energy.

Intensity is not usually measured directly but calculated from the pressure
waveform obtained by measurement of pressure using a hydrophone. In plane
waves, intensity (I) is related to the square of acoustic pressure (p) with the relation:

$$I = \frac{p^2}{\rho c} \tag{2}$$

where ρ is the density and c is the propagation speed of ultrasound in the medium.

During the passage of an ultrasound pulse through the medium, the pressure in a
certain point within the ultrasound beam varies with time (Fig. 2). Since the
intensity is related to the square of pressure, its value is always positive. The
maximum or peak value of intensity during the pulse is called the temporal peak
intensity I_{TP}. An alternative and more widely used measure of intensity during the
pulse is the pulse average intensity, I_{PA}.

The intensity also varies with position in the beam. It is possible to specify
intensity at the position in the beam where it has maximum value. This is called the
spatial-peak value. Alternatively, it is possible to calculate a value averaged over
the beam cross-sectional area, and this value is called the spatial average value. The
most often quoted intensity parameters are the following:

I_{SATA} (Spatial-average temporal-average intensity): The temporal average intensity averaged across the beam cross-section

I_{SPTA} (Spatial-peak temporal-average intensity): The temporal average intensity measured at the location where it is the largest

I_{SPPA} (Spatial-peak pulse-average intensity): The pulse average intensity measured at the location where it is the largest.

3.5   Free-Field and Derated Values

Measurements of acoustic output parameters for ultrasound medical devices are normally measured in water using hydrophones. The values of output parameters obtained in this way are commonly called free-field values. Although everyone agrees that it is good practice to use water as the standard measuring medium, it is not always representative of the real clinical situations. In order to estimate pressure values that might exist in soft tissue in the same ultrasound beam the pressure values are derated, by an amount that depends on the attenuation of the tissue. Derated values are calculated assuming that the attenuation coefficient of soft tissue is 0.3 dB $(MHz\ cm)^{-1}$. This particular value of the attenuation coefficient was chosen to be representative of typical low-loss tissue, and this so-called "derating factor" has been used widely for calculations relating to safety. Consequently, all output parameters of the ultrasonic fields which are "derated" have index 0.3.

3.6   The Purpose of Acoustic Output Measurements

Over the five decades in which diagnostic ultrasound has been in use, the magnitudes of average intensity and other acoustic output quantities have increased considerably. The need to measure acoustic output was identified very early [8, 9, 10], and values of intensities reported were of the order of the few mW cm^{-2} or few tens of mW cm^{-2} for B-mode, and up to a few hundred mW cm^{-2} for Doppler devices. Reported values of acoustic outputs increased during the 1980s [11] and 1990s [12, 13]. It was demonstrated that both time-averaged intensities and peak pressures have increased considerably. It was also found that B-mode produces the lowest spatial-peak time-averaged intensities I_{SPTA} values because the beam is scanned across a region of tissue. The highest I_{SPTA} values are produced in pulsed Doppler mode, while Colour Doppler modes tend to have I_{SPTA} values intermediate between those of pulsed Doppler and B-mode. Surveys since 1991 demonstrate that peak pressures have increased steadily. The spatial-peak temporal-average intensity (I_{SPTA}) values in B-mode have shown the greatest increase and now overlap with the range of pulsed Doppler values [13, 14].

In order to protect the public against inappropriate exposure to ultrasound when used for medical applications international standards and national regulations are developed. Individuals are commonly exposed to ultrasound as patients for diagnostic, therapeutic and surgical purposes. IEC standards for safety and essential performance of all medical electrical equipment are set out in the 60601 series. There are three particular standards in the 60601 safety series concerning medical ultrasound. These are part 2–5 for ultrasound physiotherapy equipment, part 2–37 for ultrasound diagnostic and monitoring equipment and part 2–62 for basic safety and essential performance of high intensity therapeutic ultrasound (HITU) equipment.

4 Protection Standards for Diagnostic Medical Ultrasound Devices

Although the upper limits of the output ultrasonic parameters for diagnostic ultrasound are not defined there is an obligation for equipment manufacturers to provide information about certain acoustic output parameters. Separate regulations regarding this exist in Europe and in the USA.

For example, in the USA, before an ultrasound scanner can be sold, approval must be sought from the Food and Drug Administration (FDA), and its regulations known as 510(k) [15] impose some limits. These regulations had a strong influence in setting and controlling ultrasound output levels. Elsewhere, government departments have established similar legislative processes which use similar safety criteria for allowing market approval.

In the IEC standard for diagnostic medical ultrasonic equipment, there is no upper limit on output intensity, nor any other output exposure quantity. Instead of these an International Electrotechnical Commission standard IEC 61157 requires certain acoustic output parameters to be declared if they exceed certain threshold values. Those parameters are: maximum negative or rarefaction pressure (p_r), output beam intensity (I_{OB}) and time-average intensity (I_{SPTA}). If any of the three above mentioned parameters exceeds the limit value, manufacturers must provide specific information relating to the acoustic output of their scanners in water under conditions which produce the maximum temporal-average intensity and maximum negative pressure for each mode of operation. In addition to the three parameters mentioned above manufacturers are obligated to declare total acoustic power, frequency, −6 dB beam width where the pulse pressure squared integral (PPSI) is the largest, and the mode and control setting that give maximum acoustic output values. The threshold value for maximum negative pressure is 1 MPa; for time-average intensity it's 100 mW cm^{-2} and for output beam intensity it's 20 mW cm^{-2}. If ultrasound equipment has lower values of acoustic output parameters manufacturers are not obligated to provide such detailed output information.

The IEC 606012-37 standard for ultrasound diagnostic and monitoring equipment specifies how a user shall be informed about potential hazard, through displayed indices, which are related to exposure and safety. These indices were developed in the USA in the early 1990s, and first were defined in the so-called

"Output Display Standard" (ODS) and published jointly by the American Institute of Ultrasound in Medicine (AIUM) and the National Electrical Manufacturers Association (NEMA) [16].

When ultrasound propagates through a medium, it induces a series of compressions and rarefactions which can produce various physical and chemical effects such as shear forces, acoustic streaming, cavitation, temperature and pressure changes, and radical formation. As heating and cavitation are two main damage mechanisms which are generally considered, the standard defines two safety indices, the Thermal Index (TI) and the Mechanical Index (MI). This standard requires display of safety indices only if the ultrasound scanner is capable of producing a value for either MI or TI greater than 1.0 under any operating condition. Index value must be displayed whenever it exceeds 0.4. However, the standard allows manufacturers of low output devices to avoid the display of safety indices.

4.1 The Safety Indices

The thermal index is a relative indicator of thermal risk during an ultrasound examination. Heating occurs due to absorption of ultrasound energy by tissue. How much heat could be produced when ultrasound passes through tissue depends on the power and intensity of the ultrasound and type of tissue as well. Fluids are low attenuators and would not be significantly heated as ultrasound passes through them while bone is known to have a larger attenuation coefficient. As a consequence, heating would be minimal in fluids and maximal in adult bone. The ultrasound beam width is another important parameter because it gives information about the area over which the interaction of the beam with tissue takes place.

TI is defined as:

$$TI = \frac{W}{W_{\circ}}$$

where W_{deg} is the acoustic power required to achieve an increase in temperature of 1 °C in tissue and W is the current acoustic power. As temperature rise strongly depends on the type of the tissue different versions of TI are defined. The formulae for TI can be divided into two types, those intended to predict temperature rise in soft tissue (Soft Tissue Thermal Index—TIS), and those for bone (Bone at focus thermal index (TIB) and Cranial bone thermal index (TIC).

The formula for TIS assumes a uniform homogenous medium with an attenuation coefficient of 0.3 dB cm^{-1} MHz^{-1} and is intended to predict temperature rise in soft tissue. For calculation of soft tissue heating the formula has the form TIS = AfW, where f is acoustic frequency, and W is acoustic power while constant A takes defined values for the specific scan mode and geometry. Thermal index for bone does not depend on frequency but it depends on the place where the temperature rise is estimated. TIB applies when bone is situated near the focus of an ultrasound beam, while TIC is used to predict temperature rise in bone close to the probe, as in transcranial scanning.

For each of the three thermal indices, two conditions of ultrasound exposure can be considered, giving 6 different formulae in all. Conditions of ultrasound exposure are: scanned and un-scanned exposure. These tissue model parameters differ from those of real tissue, so the TI formulation does not provide accurate measures but presents a relative measure of risk [17].

Mechanical index (MI) is a relative indicator of the possibility of occurrence of mechanical damage in the tissue as a result of internal cavitation during ultrasound exposure. Acoustic cavitation is the formation and collapse of gaseous and vapour bubbles in a liquid due to an acoustic pressure field. Stable cavitation (non-inertial) is represented by relatively long-lived gas bubbles which results in emissions at subharmonics of the main excitation frequency and can induce bubble-associated microstreaming while transient cavitation (inertial) bubbles exist for a very short period of time and collapse violently. The implosion of cavitating bubbles can generate extreme temperature and pressure conditions under which highly reactive radicals can be generated. The ability of ultrasound to cause cavitation depends on frequency and intensity of ultrasound, medium properties (viscosity and surface tension), and ambient conditions (temperature and pressure).

The formula for MI calculation is based on a mathematical model which assumes the presence of bubble nuclei in the tissue [18]. It predicts that cavitation is more likely to occur if the peak rarefaction pressure (p_r) is large and at lower frequency (f). Mechanical index (MI) is defined as:

$$MI = \frac{p_{r,0.3}}{\sqrt{f}} \tag{4}$$

where $p_{r,0.3}$ is the maximum value of derated peak rarefaction pressure (in MPa) in the beam and f is the centre frequency in the pulse in MHz. The MI is roughly proportional to the mechanical work that can be performed on a bubble in the rarefactional phase of the acoustic field. According to the theory, cavitation could not be possible at values of MI less than 0.7. This threshold applies only to bubble clouds in water while much higher values are required to initiate bubble generation in tissue in vivo [19]. Also, it is not generally possible to compare MI thresholds at different frequencies. Furthermore, application of different contrast agents will influence on MI threshold for cavitation.

The introduction of the safety indices contribute significantly to ensure safe use of ultrasound but definitions of both the TI and MI experienced much criticism. The MI and TI are output indices because their values are related to specific output parameters, but they are rather relative quantities. It is known that there are some shortcomings in the methods of calculation of indices, such as: the inappropriate assumption that the linear conditions of acoustic propagation apply and lack of consideration of the transducer as a heat source [20], etc. Another criticism has been that the index alone does not inform the user about the related depth at which the index applies. The indices are based on calculation and assumptions which lead to a maximum value of the index, irrespective of depth at which this may be found. Although the definition of both indices suffered much criticism, if they are properly used they fulfil their goal of helping to ensure patient safety. This is accomplished by application of the ALARA (as low as reasonably achievable) principle which ultrasound borrows from the application of ionizing radiation in medicine. By implementing the security index a greater responsibility is transferred on the user to limit exposures and help to protect patients from inappropriate exposure to ultrasound. Therefore, it is very important that users are well educated and informed about the meaning of safety indices and according to that appropriate use of indices in practise in order to ensure patient protection.

Even though the indices are available on screen, the user still faces the problem of knowing how long a particular tissue can be imaged. General recommendation is restricting the acoustic output to no more than what is actually required to obtain the necessary diagnostic information. In order to help users the British Medical Ultrasound Society (BMUS) has published guidance [21] which advises on actions to be taken by the user depending on the value of the displayed safety indices. It is emphasized that these are not rules to be rigidly followed but recommended guidelines only. The BMUS guidance gives recommendation for limiting exposure time at TI values higher than 0.7. For obstetric scanning the upper limit for TI is 3.0. The limit on TI of 1.0 for eye scanning is the same as that set by US regulations [15]. More detail can be found in the BMUS Detailed Guidelines.

4.2 Transducer Surface Temperature

In addition to heating due to absorption of ultrasound in the tissue, the temperature of tissue near the transducer is strongly influenced by heating of the transducer itself. Conduction of heat from the transducer surface can result in temperature rises of several degrees Celsius in the tissue close to transducer [20, 22]. Maximum allowable transducer surface temperatures are specified in IEC 60601-2-37 and these limits are presented in Table 1.

From Table 1 it can be seen that the maximum allowable transducer surface temperature is 50 °C when the transducer is transmitting in the air and 43 °C when the transducer is transmitting into tissue. Also, a higher increase in temperature is allowed if the transducer is applied externally in contact with the skin than for those used internally, such as trans-vaginal or intra-rectal probes.

Table 1 Limits on surface temperature and surface temperature rise specified by IEC 60601-2-37 (IEC 60601-2-37 [5]

	In air	On tissue (external use)	On tissue (internal use)
Maximum temperature (°C)	50	43	43
Maximum temperature rise (°C)	27	10	6

5 Quality Assurance for Ultrasound Medical Diagnostic Devices

Quality assurance (QA) programme in diagnostic radiology is an organized effort by the staff operating a facility to ensure that the diagnostic images produced are of sufficiently high quality so that they consistently provide adequate diagnostic information at the lowest possible cost and with the least possible exposure of the patient to radiation [23].

Quality control (QC) is a part of quality assurance programme consisting of observation techniques and activities to be performed in order to fulfil requirements for appropriate equipment performance.

Diagnostic ultrasound imaging is very often the basis for diagnostic decision; therefore it is necessary to implement a QA programme on such systems. Ultrasound probes degrade 10–13% per year which can, in absence of quality control, lead to wrong or no diagnosis [24].

Several recommendations have been published on diagnostic ultrasound QA [25, 26, 27, 28, 29, 30] but there is still no general agreement on a standard protocol to be followed. In 2012 EFSUMB published *Guideline for Technical Quality Assurance (TQA) of Ultrasound devices (B-Mode)* [30] with a goal to develop a standardized protocol for all EU member states. The implementation of quality assurance in some centres is still rather exception than rule. There are many possible reasons for this but the most important is probably the lack of legislative requirement. Other reasons could be: sonography workload and the lack of a medical physics department as well as general view that formal QA is unnecessary [31].

Different approaches for QA have been suggested in these publications, but the tests to be included are mostly the same in all of them. It is up to the user to define and follow their own QC protocol. Basic quality assurance tests usually includes: physical and mechanical inspection, evaluation of image uniformity, geometric accuracy, system sensitivity, dead zone, spatial resolution, grey scale evaluation and fidelity of image display. Recommended frequencies of testing and acceptability criteria are given in the Table 2.

5.1 *Physical and Mechanical Inspection*

Physical and mechanical inspection includes a visual check of the main unit, transducers, monitor, printer and accessories. Ultrasound systems and accessories, including manuals, repair records and QA records should be checked. Scanner monitor, keyboard, knobs, transducers and holders should be checked for cleanliness. Device housing, and then the air filters should be inspected and cleaned if necessary. Scanner, monitor and accessories should be properly checked, especially on portable units. Wheels should be fastened securely and the wheel locks should be working. All wires must be free of cuts and fraying to ensure proper connection.

Table 2 Frequency of quality control testing and acceptability criteria

Test	Frequency	Acceptability criteria
Physical and mechanical inspection	Daily/semi-annually	No visual defects
Image uniformity	Semi-annually	$\leq 20\%$ from baseline
Geometric accuracy: vertical	Semi-annually	≤ 2 mm or 2%
Geometric accuracy: horizontal	Semi-annually	≤ 3 mm or 3%
System sensitivity	Semi-annually	≤ 1 cm from baseline
Dead zone	Semi-annually	≤ 10 mm for f < 3 MHz ≤ 7 mm for 3 MHz $\leq$ f < 7 MHz ≤ 4 mm for f $\geq$ 7 MHz
Spatial resolution: axial	Semi-annually	≤ 1 mm
Spatial resolution: lateral	Semi-annually	≤ 1.5 mm from baseline
Grey scale evaluation	Semi-annually	$\leq 10\%$ from baseline

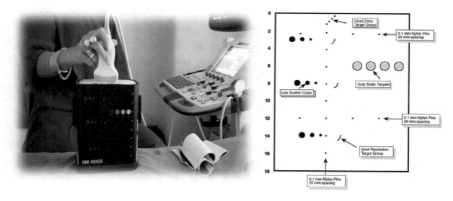

Fig. 3 Example of a general-purpose image quality phantom

5.2 Phantom Based Measurements

Phantoms for QC testing of diagnostic ultrasound are made of tissue mimicking material with various inserts designed for a certain test. The tissue mimicking material within the phantom consists of water based gelatine in which microscopic particles are mixed uniformly through the volume. The speed of sound in this material is about 1540 m/s, as the same speed is assumed in the calibration of ultrasound scanners. There are several manufacturers providing such phantoms and they are very similar to one another. An example of a general-purpose quality assurance phantom in use is shown in Fig. 3. For analysis either visual or automatic methods can be used, but according to the guidelines of the European Federation of Societies for Ultrasound in Medicine and Biology (EFSUMB) automatic image analysis is recommended [30]. When setting up baseline values, scanner presets for each test should be recorded and reproduced in future periodical testing.

a. *Image uniformity*

Uniformity is defined as the ability of an ultrasound device to display echoes of the same magnitude and depth with equal brightness on display. This is a good test to ensure that all crystals within the transducer are functioning similarly. The phantom section for this test is the tissue mimicking material of uniform texture similar to that of liver parenchyma which is free of filament and lesions—simulating targets [27]. Uniformity could be evaluated by performing Region of Interest (ROI) measurements (Fig. 4) using an image processing programme, for example ImageJ (Research Services Branch of the National Institute of Mental Health, USA). The standard deviation and mean are measured corresponding to the echo level and noise measurements, respectively.

b. *Geometric accuracy: vertical and horizontal*

Distance accuracies, both in the direction of the ultrasound beam and perpendicular to the ultrasound beam are contributors to total image quality and are therefore indicators of scanner performance.

Vertical distance accuracy
   Vertical distance is defined along the axis of the beam. The phantom section for this test is the vertical distance target group, vertical column of filament targets equally separated (e.g. 2 cm) in different depths. The distance between the most clearly separated filament targets in the vertical column displayed in the image is measured using the distance tool in the image processing programme (Fig. 5).

Horizontal distance accuracy
   Horizontal distance is defined perpendicular to the axis of the beam. The phantom section for this test is the horizontal distance target group, horizontal rows of filament targets equally separated (e.g. 3 cm) in different depths. The distance

Fig. 4 Image for uniformity evaluation. Red circles represent Regions of Interest (ROIs) for the analysis of echo level and noise

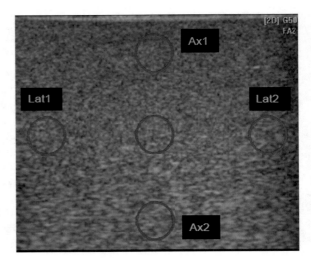

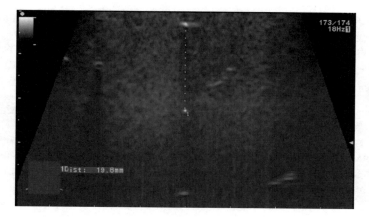

Fig. 5 Vertical distance measurement using distance tool

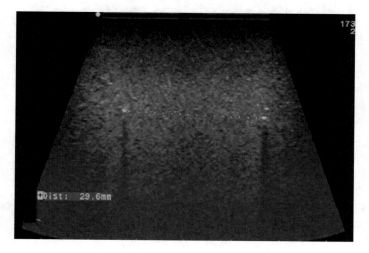

Fig. 6 Horizontal distance measurement using distance tool

between the most clearly separated filament targets in the horizontal row displayed in the image is measured using the distance tool in the image processing programme (Fig. 6).

c. *System sensitivity*

System sensitivity is described by the depth of penetration. This is the largest distance in the phantom for which echo signals from the scatters within the tissue mimicking background material can be detected on the display. The frequency of the transducer, the attenuation of the medium being imaged and the system settings determine the depth of penetration.

The phantom section for this test has the vertical group of filaments, in this case "depth markers" that are equally spaced (e.g. 2 cm) one from another. The point at which the usable tissue information disappears is a good indicator of maximum depth of visualization or penetration (Fig. 7).

d. *Dead zone*

The ring down or dead zone is the distance from the front face of the transducer to the first identifiable echo. No useful scan data are collected in this region. The dead zone is the result of the transducer ringing and reverberations from the transducer-test object, phantom or patient interface [27]. The phantom section for this test is the one containing pin targets in close proximity to the phantom surface at different shallow depths (e.g. 1, 4, 7, and 10 mm) (Fig. 8).

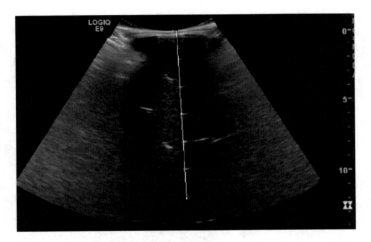

Fig. 7 Measurement of the depth of visualization using distance tool which describes the system sensitivity

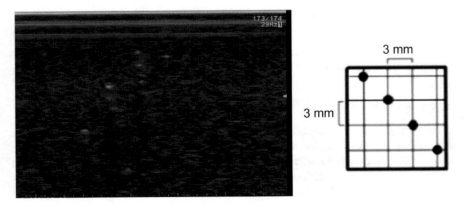

Fig. 8 Image for the visual evaluation of the dead zone and an illustration of the dead zone insert

e. *Spatial resolution: Axial and lateral*

The axial resolution is defined as the ability of an ultrasound system to resolve objects in close proximity along the axis of the beam. Axial resolution is proportional to the probe frequency and is approximately double the pulse length. Lateral resolution is a measure of how close two reflectors can be to one another, perpendicular to the beam axis and still be distinguished as separate reflectors. It is defined in the direction perpendicular to the beam axis. Lateral resolution is determined by the width of the ultrasound beam and therefore is detected to be at its best within the focal zone.

The phantom usually has three sets of resolution target groups at different depths (e.g. 3, 8 and 13 cm). The target consists of four wires of small diameter vertically and horizontally spaced at known distances (e.g. vertically 2, 1, 0.5 and 0.25 mm and horizontally 1 mm) (Fig. 9). Both axial and lateral resolution is evaluated using the same group. Axial resolution corresponds to the last pair of points identifiable as two separate entities in the target group. Lateral resolution corresponds to the width of a single filament in the focal zone.

f. *Grey scale evaluation*

Grey scale evaluation of an ultrasound image is a very important parameter because clinical differentiation of the various lesions in tissue depends greatly on the grey scale characteristic of the obtained image. The phantom section utilized for this test consists of a set of four spherical objects each having the same diameter (e.g. 1 cm), placed at the same depth (e.g. 6 cm) with different contrast (e.g. 0, −6, +6 and + 12 dB) (Fig. 10). For the quantitative measurement the mean pixel value is computed using the ROI tool.

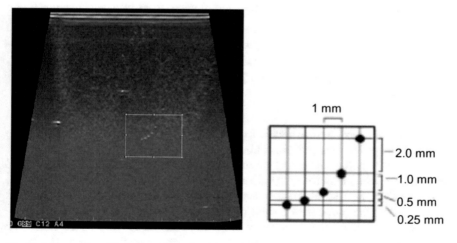

Fig. 9 Image and an illustration of the spatial resolution target group

Fig. 10 Image of a grey
scale evaluation insert; four
spherical objects of different
contrast

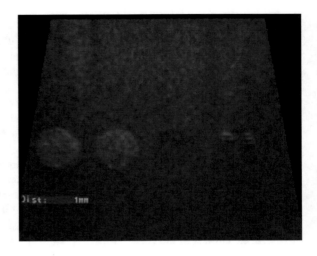

5.3 Fidelity of Ultrasound Scanner Electronic Image Display

Ultrasound images are viewed on electronic image displays. Therefore, they must
be appropriately set up and monitored. Many ultrasound units have "built-in" or
internally stored grey scale test patterns that can be used for display setup. Less
comprehensive variation of the protocol for QC of diagnostic image display in
general can be used. Testing of image displays is in accordance with the AAPM
TG18 report [32] and is using TG18 test patterns (Fig. 11). Such patterns should
not be the only tool for evaluation. Clinical images should also be examined.
Setting up the displays for the first time, to establish the operating levels, is best
done with the advice of one or more radiologists to ensure that the levels are
clinically acceptable.

6 Protection Standards for Ultrasound Physiotherapy Equipment

Ultrasound has been used as a therapeutic technique for physical medicine since the
1950s. The need to measure and calibrate physiotherapy ultrasound equipment was
identified in early the 1960s, and specification standards were first published by IEC
in 1963. Requirements for physiotherapy machines are small in number and well
defined [33]. The IEC 60601-2-5 standard for physiotherapy equipment includes
two limits for the purpose of patient protection.

The first one limits the temperature of the front face of the transducer. It must be
less than 41 °C when operated under water with initial temperature 25 °C.

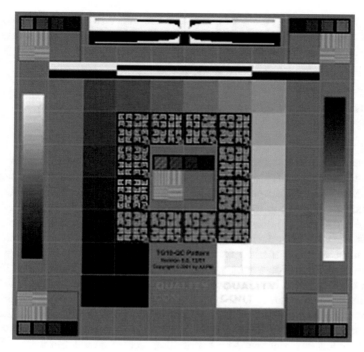

Fig. 11 AAPM TG18-QC pattern used for overall evaluation of image display performance

According to the standard the equipment must operate for 3 min at the maximum effective ultrasound output power. The treatment head is then removed from water for 15 s and then immediately re-immersed in the water. The above procedure should be repeated two more times.

The second protection limit applies to ultrasound intensity. According to IEC standard 60601-2-5 the effective intensity shall not exceed 3 W/cm^2. In the same document the effective intensity is defined as the quotient of the effective ultrasound power and the effective radiating area (ERA).

The ultrasonic beam distribution generated by the treatment head is another important parameter concerning safety. Beam homogeneity can be quantified by the parameter called beam non-uniformity ratio (BNR), which represents the ratio of the highest intensity in the field to the intensity averaged over the effective radiating area. Sometimes, beam distribution can be non-uniform and can potentially generate regions of high local pressure, also called "hot spots" [34]. These regions may produce excessive heating in small regions of the tissue. According to IEC standard 61689 [35] transducers with BNR > 8 are considered unsafe.

Since the first introduction of standards and limits for output parameters of the devices used in physiotherapy, numerous surveys have been published [36, 37]. These papers show that lot of physiotherapy devices have significant differences between the indicated and actual output power. Also, some cases of excessive heating as a result of equipment failure have been cited in literature [38]. All these publications have emphasized the need for measurement and calibration of the equipment used in physiotherapy.

6.1   Measurement of Ultrasonic Power for Physiotherapy Devices

Ultrasound transducer output power is a key safety related parameter. The clinical effects of the ultrasound physiotherapy depend on applied acoustic dose defined as energy deposited by absorption of acoustic wave per unit mass of the medium [39]. If the acoustic dose is too low there will be no significant clinical effects and if it is too large it can cause tissue injury [40]. Therefore, it is very important to measure and calibrate ultrasonic devices used in therapy, in order to preserve patient safety.

The most common method for determining ultrasound power includes a measurement of the radiation force by using radiation force balances (RFB). The basic concept of the method has been previously explained. This method is based on the fact that ultrasound exerts a force on a target placed in the path of an acoustic beam that is directly proportional to the total power absorbed or reflected by the target. The small magnitude of the radiation force makes the method difficult to apply within a clinical environment. Another limiting factor for radiation force balances is that it can register only the component of the radiation force which is collinear with the axis of measurement. This fact may cause problems in the measurement of power for divergent or convergent beams.

There are some alternative methods for measurement of acoustic power based on the conversion of acoustic energy into thermal energy such as calorimetry [41] and pyroelectricity [42].

The calorimetric method for measurement of acoustic power is attractive because the principle of measurement is simple. The absorption of ultrasound energy in an absorbing liquid such as castor oil produces a temperature rise. The acoustic power may be then calculated from the resultant increase in temperature. Disadvantages of these methods are that it is time consuming and has low sensitivity. Also, this method will give a good result for acoustic power only if all acoustic energy is converted into thermal energy, and if thermal losses can be ignored. Advantages of the calorimetric method are that it is insensitive to the direction of ultrasound propagation and can be used for measuring power of divergent and convergent beams [33].

Recently, the UK National Physical Laboratory (NPL) has developed a new method for measurement of acoustic power which exploits the pyroelectric effect generated within a thin layer of the piezo-polymer polyvinylidene fluoride (PVDF). One side of the membrane is bonded to a very highly acoustically absorbing material. The majority of the ultrasonic energy passing through the PVDF membrane is absorbed in the thin layer of backing material, producing a rapid increase in temperature and generation of pyroelectric voltage which is proportional to delivered ultrasonic power. The measurement concept was introduced in the paper by [43], with follow up publications addressing the measurement of physiotherapy fields [42] and of diagnostic fields [44, 45].

6.2 Determination of Effective Radiating Area

The effective radiation area (ERA) is the area close to the face of the transducer over which the majority of the ultrasonic power is emitted. Accurate determination of the effective radiation area of the transducer is particularly important because it effects the calculation of the effective intensity which should not exceed 3 W/cm^2 as mentioned previously. The geometric surface of the piezoelectric crystal itself is not a reliable value for the area over which ultrasound is emitted. Also, ultrasound physiotherapy treatment heads require periodic checks as their performance tends to deteriorate slowly, mainly resulting from minor damage to the transducer probe. Hence, accurate determination of ERA is of particular importance.

Performance of the treatment head can be checked and determination of effective radiating area can be performed using the standardized method which is based on the scanning of the acoustic field in water tanks using miniature hydrophones. These procedures require a specially equipped laboratory and they are not very convenient for use in clinical environment. Recently, an alternative method based on the use of thermochromics materials was proposed [46] and tested in a clinical environment [47]. This method is useful for rapid quality assurance of physiotherapy ultrasound treatment heads.

There are two specification standards which describe methods for determination of the ERA. According to standard IEC 61689, ERA is evaluated through the derivation of an intermediate quantity called the beam cross-sectional area (BCSA). The BCSA is defined as the minimum area which contains 75% of the total mean square acoustic pressure. It is determined by sorting analysis of the acquired data. ERA of the treatment head is calculated by multiplying the beam cross-sectional area determined at a distance of 0.3 cm from the treatment head's face, by a dimensionless factor F_{AC}. This method allows the determination of real values for ERA.

Prior to the publication of IEC 61689, the majority of manufacturers measured ERA using technique specified by the FDA. This method involved measuring the maximum pressure in the beam and then defining an area within which pressure amplitude exceed 5% of this maximum. This method can result in large measurement uncertainties.

References

1. Manbachi A, Cobbold RSC (2011) Development and application of piezoelectric materials for ultrasound generation and detection. Ultrasound 19(4):187
2. Kremkau FW (2006) Diagnostic ultrasound. Principles and instruments, Chap. 4, 7th edn. Saunders, Philadelphia, USA
3. Macovski A (1983) Medical imaging systems, Chap. 10. Prentice-Hall, Englewood Cliffs, NJ
4. IEC 60601 part 2–5, Medical electrical equipment: particular requirements for the safety of ultrasound physiotherapy equipment. International Electrotechnical Commission, Geneva
5. IEC 60601 part 2–37: Medical electrical equipment: particular requirements for the safety of ultrasound diagnostic and monitoring equipment, 2001 consolidated with amendment 1, 2004. International Electrotechnical Commission, Geneva
6. IEC 60601 part 2–62 (2013) Medical electrical equipment: particular requirements for the basic safety and essential performance of high intensity therapeutic ultrasound (HITU) equipment. International Electrotechnical Commission, Geneva
7. Perkins MA (1989) A versatile force balance for ultrasound power measurement. Phys Med Biol 34(11):1645
8. Farmery MJ, Whittingham TA (1978) A portable radiation-force balance for use with diagnostic ultrasonic equipment. Ultrasound Med Biol 3(4):373–379
9. Carson PL, Fischella PR, Oughton TV (1978) Ultrasonic power and intensities produced by diagnostic ultrasound equipment. Ultrasound Med Biol 3(4):341–350
10. Farmery MJ, Whittingham TA (1978) A portable radiation-force balance for use with diagnostic ultrasonic equipment. Ultrasound Med Biol 3:373–379
11. Duck FA (1989) Output data from European studies. Ultrasound Med Biol 15(Suppl 1):61–64
12. Duck FA, Martin K (1991) Trends in diagnostic ultrasound exposure. Phys Med Biol 36 (11):1423–1432
13. Henderson J, Willson K, Jago JR, Whittingham TA (1995) A survey of the acoustic outputs of diagnostic ultrasound equipment in current clinical use. Ultrasound Med Biol 21(5):699–705
14. Whittingham TA (2000) The acoustic output of diagnostic machines. In: ter Haar G, Duck FA (eds) The safe use of ultrasound in medicine diagnosis, 2nd edn. BMUS/BIR, London, UK
15. FDA (1997) Information for manufacturers seeking marketing clearance of diagnostic ultrasound systems and transducers. Division of Reproductive, Abdominal, Ear, Nose, Throat and Radiological Devices: Office of Device Evaluation. US Department of Health and Human Services: Food and Drug Administration
16. AIUM/NEMA (1992) UD2: Acoustic output measurement standard for diagonstic equipement (Rockville, MD: American Institute of Ultrasound in Medicine)
17. Bigelow TA, Church CC, Sandstrom K, Abbott JG, Ziskin MC, Edmonds PD, Herman B, Thomenius KE, Teo TJ (2011) The thermal index: its strengths, weaknesses, and proposed improvements. J Ultrasound Med 30(5):714–734
18. Apfel RE, Holland CK (1991) Gauging the likelihood of cavitation from short-pulse, low-duty cycle diagnostic ultrasound. Ultrasound Med Biol 17(2):179–185
19. Curch CC (2002) Spontaneous homogeneous nucleation, inertial cavitation and the safety of diagnostic ultrasound. Ultrasound Med Biol 28(10):1349–1364
20. Duck FA, Starritt HC, ter Haar GR, Lunt MJ (1989) Surface heating of diagnostic ultrasound transducers. Br J Radiol 62:1005–1013
21. BMUS (2010) Guidelines for the safe use of diagnostic ultrasound equipment. Ultrasound 18:52–59
22. Calvert J, Duck F, Clift S, Azaime H (2007) Surface heating by transvaginal transducers. Ultrasound Obstet Gynecol 29:427–432
23. Quality assurance in diagnostic radiology: a guide prepared following a workshop held in Neuherberg, Federal Republic of Germany, 20–24 October 1980, and organized jointly by Institute of Radiation Hygiene, Federal Health Office, Neuherberg, Federal Republic of

Germany, Society for Radiation and Environmental Research, Neuherberg, *Federal Republic of Germany and World Health Organization*, Geneva, Switzerland, 1982

24. Hangiandreou NJ, Stekel SF, Tradup DJ, Gorny KR, King DM (2011) Four-year experience with a clinical ultrasound quality control program. Ultrasound Med Biol 37(8):1350

25. American College of Radiology (ACR) (2011) ACR technical standard for diagnostic medical physics performance monitoring of real time ultrasound equipment. ACR, Reston, VA

26. American Institute of Ultrasound in Medicine (AIUM) (2008) Routine quality assurance for diagnostic ultrasound equipment. AIUM, Laurel, MD

27. Goodsitt MM, Carson PL, Witt S, Hykes DL, Kofler JM (1998) Real-time B-mode ultrasound quality control test procedures. Report of AAPM Ultrasound Task Group No. 1. Med Phys 25:1385–1406

28. International Electrotechnical Commission, IEC 62736-1 (2011) Ultrasonics—quality control of diagnostic medical ultrasound systems

29. Institute of Physics and Engineering in Medicine (IPEM) (2010) Quality assurance of ultrasound imaging systems, Report No 102, IPEM, York

30. Kollmann C, Dekorte C, Dudley NJ, Gritzmann N, Martin K, Evans DH (2012) Guideline for technical quality assurance (TQA) of ultrasoud devices (B-Mode)—Version 1.0 (July 2012). EFSUMB Technical Quality Assurance Group. Ultraschall Med 33:544–549

31. Dudley N, Russell S, Ward B, Hoskins P (2014) The BMUS guidelines for regular quality assurance testing of ultrasound scanners. Ultrasound 22:6–7

32. Samei E, Badano A, Chakraborty D, Compton K, Cornelius C, Corrigan K, Flynn MJ, Hemminger B, Hangiandreou N, Johnson J, Moxley-Stevens DM, Pavlicek W, Roehrig H, Rutz L, Shepard J, Uzenoff RA, Wang J, Willis CE (2005) Assessment of display performance for medical imaging systems: executive summary of AAPM TG18 report. Med Phys 32(4):1205–1225

33. Shaw A, Hodnett M (2008) Calibration and measurement issues for therapeutic ultrasound. Ultrasonics 48:234–252

34. Gutiérrez MI, Calás H, Ramos A, Vera A, Leija L (2012) Acoustic field modeling for physiotherapy ultrasound applicators by using approximated functions of measured non-uniform radiation distributions. Ultrasonics 52(6):767–777

35. IEC 61689 (2007) Ultrasonics—physiotherapy systems—field specifications and methods of measurement in the frequency range 0.5 MHz to 5 MHz

36. Hekkenberg RT, Oosterbaan WA, van Beekum WT (1986) Evaluation of ultrasound therapy devices. Physiotherapy 72:390–395

37. Pye SD, Milford C (1994) The performance of ultrasound physiotherapy machines in Lothian region, Scotland. Ultrasound Med Biol 20(4):347–359

38. Pye SD (1996) Ultrasound therapy equipment—does it perform? Physiotherapy 82(1):39–44

39. Duck F (2009) Acoustic dose and acoustic dose rate. Ultrasound Med Biol 35(10):1679–1685

40. Miller D, Smith N (2012) Overview of therapeutic ultrasound applications and safety considerations. J Ultrasound Med 31(4):623–634

41. Pye S, Hildersley K, Somer E, Munro V (1994) A simple calorimeter for monitoring the output power of ultrasound therapy machines. Physiotherapy 80(4):219–223

42. Zeqiri B, Barrie J (2008) Evaluation of a novel solid-state method for determining the acoustic power generated by physiotherapy ultrasound transducers. Ultrasound Med Biol 34:1513–1527

43. Zeqiri B, Gélat PN, Barrie J, Bickley CJ (2007) A novel pyroelectric method of determining ultrasonic output power: device concept, modelling and preliminary studies. IEEE Trans Ultrason Ferroelectr Freq Control 54(11):2318–2330

44. Zeqiri B, Žauhar G, Hodnett M, Barrie J (2011) Progress in developing a thermal method for measuring the output power of medical ultrasound transducers that exploits the pyroelectric effect. Ultrasonics 51:420–424

45. Zeqiri B, Žauhar G, Rajagopal S, Pounder A (2012) Systematic evaluation of a secondary method for measuring diagnostic-level medical ultrasound transducer output power based on a large-area pyroelectric sensor. Metrologia 49(3):368–381

46. Butterworth I, Barrie J, Zeqiri B, Žauhar G, Parisot B (2012) Exploiting thermochromic materials for the rapid quality assurance of physiotherapy ultrasound treatment heads. Ultrasound Med Biol 38(5):767–776
47. Žauhar G, Radojčić ĐS, Dobravac D, Jurković S (2015) Quantitative testing of physiotherapy ultrasound beam patterns within a clinical environment using a thermochromic tile. Ultrasonics 58:6–10

Inspection and Testing of Electroencephalographs, Electromyographs, and Evoked Response Equipment

Mario Cifrek

Abstract The chapter deals with the inspection of neurodiagnostic equipment based on measurement of electrophysiological signals in order to detect eventual problems and prevent them from becoming serious safety risks. In the first section of the text a brief description of the human neuromuscular system is given, followed by description and short historical overview of considered neurodiagnostic methods: electroencephalography (EEG) including evoked potentials (EP), electromyography (EMG) and nerve conduction study (NCS). Operating principle of computer-controlled neurodiagnostic instrument is explained using generic block diagram. The next sections discuss potential harms and hazards associated with the use of neurodiagnostic equipment as well as standards and regulations concerning basic safety and essential performance requirements for mentioned equipment. The last section describes inspection procedure for periodical testing of modern computer-based nerodiagnostic instruments in the field.

1 Introduction

To treat disorders of the brain, spinal cord, muscles and peripheral nervous system, neurologists use a variety of diagnostic tests to help identify the specific nature of neurological diseases, conditions, or injuries. The results of these tests can help in planning an appropriate course of treatment. Some of these tests are provided using electroencephalography (EEG)—a method for recording spontaneous as well as evoked (evoked potentials, EP) brain electrical activity, nerve conduction study (NCS) comprising action potential morphology and nerve conduction velocity (NCV) measurement and electromyography (EMG), a method for recording the electrical activity of muscles.

M. Cifrek (✉)
Faculty of Electrical Engineering and Computing, University of Zagreb,
Unska 3, Zagreb, Croatia
e-mail: mario.cifrek@fer.hr

© Springer Nature Singapore Pte Ltd. 2018
A. Badnjević et al. (eds.), *Inspection of Medical Devices*, Series in Biomedical
Engineering, https://doi.org/10.1007/978-981-10-6650-4_7

2 Human Neuromuscular System

The human neuromuscular system includes the nervous system and the muscular system. The nervous system is anatomically divided into two parts: the central nervous system (CNS) which is made up of the brain and spinal cord, and the peripheral nervous system (PNS), which consists of 12 pairs of cranial and 31 pairs of spinal nerves along with their associated ganglia [1]. Functionally, the nervous system has two main subdivisions: the autonomic (involuntary) and the somatic (voluntary) nervous system. The autonomic nervous system maintains internal physiologic homeostasis, regulating certain body processes, such as blood pressure and the rate of breathing, which work without conscious effort. The somatic system is made up of two different types of neurons: sensory neurons (afferent neurons), which transmit messages to the central nervous system and motor neurons (efferent neurons) which relay information from the central nervous system toward the peripheral effector organs, mainly muscles and glands [2].

Different parts of the brain are responsible for different tasks. A basic division of the brain is into four lobes: frontal, temporal, parietal and occipital [3]. For example, the visual cortex is situated in the occipital lobe, the primary somatosensory cortex is in the anterior part of the parietal lobe and the primary motor cortex is in the posterior part of the frontal lobe. A large fissure called the central sulcus separates the primary somatosensory and the primary motor cortex. Different parts of the primary somatosensory and motor cortex are responsible for different parts of the body [4]. The left side of the brain is mostly responsible for movements of the right side of the body and vice versa.

Nerve conduction velocity—the velocity with which a signal (an action potential) propagates through a neuron—depends on the type of the nerve fibre and spans the range from 0.5 m/s (the slowest, type C unmyelinated fibres) to 120 m/s (the fastest, type A myelinated fibres) [2]. Conduction velocities are affected by a wide array of factors, including age, sex and various medical conditions. Studies allow for better diagnoses of various neuropathies, especially demyelinating conditions as these conditions result in reduced or non-existent conduction velocities.

The muscular system is responsible for the movement of internal body parts as well as the movement of whole human body. There are three types of muscle tissue: smooth (visceral), cardiac and skeletal. Smooth muscle is found inside of organs like the stomach, intestines and blood vessels, where it makes organs contract to move substances through them. It is known as involuntary muscle because it is controlled by the unconscious part of the brain. The term "smooth muscle" is often used because it has a very smooth, uniform appearance when viewed under a microscope. This uniform appearance contrasts with the banded appearance of cardiac and skeletal muscles. Found only in the heart, cardiac muscle is responsible for pumping blood throughout the body. Cardiac muscle tissue cannot be controlled consciously, so it is an involuntary muscle. Skeletal muscle, also known as the striated muscle, is the only voluntary muscle tissue in the human body. It is required for every physical action that a person performs consciously (e.g. speaking, walking, or writing). The term "striated muscle" is used because the banded appearance observed in microscopic images of this tissue.

Skeletal muscles are innervated by alpha-motoneurons, which have heavily myelinated, fast-conducting axons that terminate in motor end plates (neuromuscular junction). In clinical neurology, for motor neurons that innervate the voluntary muscles, the term "lower motoneuron" is used. A single alpha motoneuron and all the muscle fibres it innervates is called a motor unit. The number of muscle fibres within motor units varies from 3 to 8 muscle fibres in the small finely controlled extraocular muscles of the eye to as many as 2000 muscle fibres in postural muscles, for example soleus muscle in the leg. Individual muscle fibres are innervated by neuromuscular junction, usually located near the middle of the cell. The muscle fibres belonging to a motor unit are dispersed and intermingle with those from other motor units.

There are three functional types of muscle fibre: (1) slow-twitch fibres, (2) fast-fatigable twitch fibres and (3) fast-fatigue-resistant twitch fibres. Each motor unit comprises only one type. The force that slow-twitch fibres produce in response to an action potential rises and falls slowly then force produced by fast-twitch fibres. The fatigue resistance of slow-twitch fibres results from a reliance on oxidation catabolism in contrast with one subtype of the fast-fatigable-twitch fibres that relies almost exclusively on anaerobic catabolism [5].

3 Diagnostic Methods and Equipment

3.1 Electroencephalography (EEG)

Electroencephalography (EEG) is an electrophysiological method to record electrical activity of the brain. The EEG is used in the evaluation of brain disorders. Most commonly it is used to show the type and location of the activity in the brain during a seizure. It also is used to evaluate people having problems associated with brain function including confusion, coma, tumors, long-term difficulties with thinking or memory or weakening of specific parts of the body (such as weakness associated with a stroke).

A thorough history of electroencephalography is given in the book by Niedermeyer and Da Silva [6], while the brief history is summarized in Table 1 [7].

Electrical activity of the brain can be measured and recorded non-invasively via surface electrodes mounted on the scalp (EEG) or invasively, directly from the surface of the brain (electrocorticography, ECoG). The amplitude of the EEG is from a few μV to 100 μV when measured on the scalp. The bandwidth of this signal is from under 1 Hz to about 50 Hz. ECoG involves recording electrical signals from the surface of the human brain, typically in patients being monitored prior to surgery. ECoG signals has a much higher signal-to-noise ratio (SNR) than EEG, as well as higher spectral and spatial resolution.

Table 1 A brief history of EEG [7]

1928–1938	First recording of human EEG from the scalp by Berger, Adrian and Matthews
1937–1945	Studies by Grey Walter, Gibbs, Gibbs and Lennox show changes related to epilepsy and when the investigation may be used clinically
1950s–1970s	8 or 16 channel hard-wired recordings on to paper become incorporated into routine clinical practice
1970s	First use of CCTV linked to analogue EEG
1980s	Analogue 3 or 4 channel ambulatory EEG introduced for long-term monitoring
1990s–2005	Digital EEG gradually replaces analogue recording
2005–current	Digital EEG with simultaneous video for both standard recording and telemetry, and ambulatory monitoring

Non-invasive EEG uses cup electrodes made of stainless steel, tin, gold or chloride treated silver disks 4–10 mm in diameter placed on the scalp surface, and clip electrodes for the earlobes typically used as reference electrode. Most used are silver/silver chloride electrodes because they minimize the contact resistance between electrode and skin surface and guarantee a stable, reliable contact. By means of a special conductive gel, which is applied in the hair between electrodes and scalp, the electrical resistance can be minimized.

Invasive EEG uses subdermal "needle" electrodes, multipole depth electrodes designed for introduction directly into the substance of the brain by a neurosurgeon, and corkscrew needle electrode. Cortical electrodes are used for recording directly from the surface of the brain (electrocorticography, ECoG). A typical grid has a set of 8 × 8 platinum-iridium electrodes of 4 mm diameter (2.3 mm exposed surface) embedded in silicon with an inter-electrode distance of 1 cm.

Surface electrodes can be placed on the scalp individually using some sort of adhesive paste or they can be placed using an elastic cap with electrode holders placed at prearranged positions. Electrode caps can be made for up to 256 electrodes. Using a very large number of electrodes does not result in a much better spatial resolution, so in most cases the number of electrodes is limited to 20–30 electrodes. For some clinical applications, just a few electrodes are used. The most frequently used system includes 32 electrodes placed according to the International 10–20 system [8]. The placement of electrodes is based on landmarks on the skull, namely the nasion (Nz), the inion (Iz) and the left and right pre-auricular points (LPA and RPA). Each electrode site is labelled with a letter and a number. The letter refers to the area of brain underlying the electrode e.g. F—Frontal lobe, T—Temporal lobe, P—Parietal lobe, O—Occipital lobe, C—border between frontal and parietal lobe. Even numbers denote the right side of the head and odd numbers the left side of the head. The electrodes that has the letter Z (for zero) as their index instead of a number are positioned at the middle of the scalp. This placement system was devised in order to get comparable data between different laboratories, participants, and experiments. In addition, the system makes sure that all regions of the brain are sampled evenly.

Besides evaluation of spontaneous electrical activity of the brain, EEG is used for evoked potential (EP) measurements. Evoked potential is an electrical potential recorded from the nervous system following presentation of a stimulus, for example, electrical, visual or auditory. Signals can be recorded from cerebral cortex, brain stem, spinal cord and peripheral nerves. Usually the term "evoked potential" is reserved for responses involving recording from central nervous system structures. Thus, evoked compound motor action potentials (CMAP) or sensory nerve action potentials (SNAP) as used in nerve conduction studies (NCS) are generally not thought of as evoked potentials, though they do meet the above definition. Examples of evoked potentials are visual evoked potential (VEP), somatosensory evoked potential (SSEP), auditory evoked potential (AEP) and auditory brainstem response (ABR). Evoked potential signals are usually below the noise level and thus not readily distinguished and a train of stimuli and signal averaging must be used to improve the signal-to-noise ratio.

Along with visual analysis of raw data, modern EEG systems enables quantitative analysis of the digitized EEG (quantitative EEG, qEEG). The qEEG is an extension of the analysis of the visual EEG interpretation which may assist and even augment our understanding of the EEG and brain function. One example of qEEG is brain mapping.

3.2 Electromyography (EMG)

Electromyography (EMG) is an electrodiagnostic method for evaluation and recording of the electrical activity of muscle tissue, or its representation as a visual display or audible signal, using electrodes attached to the skin or inserted into the muscle. EMG is performed using an instrument called an electromyograph to produce a record called an electromyogram. According to IEC 60601-2-40, electromyograph is medical electrical equipment for the detection or recording of biopotentials accompanying nerve and muscle action, either spontaneously, intentionally or evoked by electrical or other stimulation.

EMG can detect abnormal muscle electrical activity in many diseases and conditions, including inflammation of muscles, pinched nerves, damage to nerves in the arms and legs, disc herniation and degenerative diseases such as muscular dystrophy, Lou Gehrig's disease and Myasthenia gravis, among others. The EMG helps distinguish between muscle conditions that begin in the muscle and nerve disorders that cause muscle weakness. The results of this test are often correlated with the results from the Nerve Conduction Study.

After the first experiments dealing with connection between muscles and the generation of electricity by Francesco Redi in 1666, Emil du Bois-Reymond in 1849 discovered that it was possible to record electrical activity during a voluntary muscle contraction. The first actual recording of this activity was made by Marey in 1890, who also introduced the term electromyography. In 1922, Gasser and Erlanger used an oscilloscope to show the electrical signals from muscles [9].

Table 2 A brief history of electromyographs

1942	The first electromyograph was constructed by Herbert Jasper in 1942 at McGill University, Montreal Neurological Institute
1950	DISA A/S (Denmark) introduced in 1950 first commercially available three channel EMG system (model 13A67)
1950–1973	The era of the analogue EMG systems: EMG signals were recorded, and subsequent analyses were carried out manually on film or paper
1973–1982	The first modular digital EMG systems were introduced. Dedicated analysis modules were introduced, but detailed analysis was still done on paper
1982–1993	Microprocessor-controlled EMG systems
1991–2005	PC-based EMG systems
2001–current	Handheld and wireless EMG systems

The first application of clinical electromyography was presented in the paper of Buchtal and Clemmesen [10]. A thorough history of electromyography is given in the book by Medved [9], while the brief history of electromyographs is summarised in Table 2 [11].

There are two types of EMG: intramuscular (needle and fine-wire) EMG and surface EMG. For intramuscular EEG, when the position of the electrode is adjusted so as to give a maximal single fibre action potential, the recorded amplitude is usually between 1 and 7 mV with occasional values of 15–20 mV. Most of the spectral energy is concentrated between 100 and 10 kHz, with a peak value at 1.6 kHz [12]. The amplitude of surface EMG signal is less than 5 mV and the bandwidth of this signal is from about 20 to 500 Hz [13].

By the intramuscular EMG muscle function is commonly studied with concentric or monopolar needle electrodes. The recordings are performed with the muscle at rest, during slight voluntary contraction and during increasing or full contraction. The types of EMG studies include: spontaneous activity (SPA), maximum voluntary activity (MVA), automatic motor unit potential analysis (AMUP), interference pattern analysis (IPA), quantitative EMG (QEMG), and single fibre EMG (SFEMG).

Surface EMG (sEMG) involves placing electrodes on the skin over the muscle to detect the electrical activity of the muscle. It is used in kinesiological studies (gait and movement analysis), fatigue studies [14] mapping of the end-plate area of a muscle [15], and biofeedback devices [16].

3.3 Nerve Conduction Study (NCS)

A nerve conduction study (NCS) is a diagnostic test used to evaluate the electrical conduction of the motor and sensory nerves in the human body. Both the nerve conduction velocity (NCV) and action potential morphology are analyzed. Nerve conduction studies are used mainly for evaluation of paresthesia's and/or weakness of the arms and legs. Nerve conduction studies along with needle electromyography measure nerve and muscle function and may be indicated when there is pain in the limbs, weakness from spinal nerve compression, or concern about some other neurologic injury or disorder.

Although the first actual recordings of afferent nerve conduction velocity was made by Eichler in 1938 [17] the work of Dawson and Scott led to the clinical development of these methods. They use the high-gain, low-noise amplifiers and a technique of photographic superimposition for recording sensory action potentials [18]. Based on the methods of Dawson and Scott, Gilliatt and Sears pioneered the use of nerve conduction studies in clinical practice. In 1955, they set up a routine recording laboratory, equipped mostly with instruments constructed by Bert Morton [19]. In 1965 they introduce a barrier grid storage tube as an "averager" [20]. In 1965 Data Laboratories Ltd. made "Biomac 500", one of the first medical averaging computer.

NCS are done by placing electrodes on the skin and stimulating the nerves through electrical impulses. To study motor nerves, recording electrodes are placed over a muscle that receives its innervation from the stimulated nerve. The electrical response of the muscle is then recorded in order to determine how fast and how well the nerve responded. NCS can be performed on any accessible nerve including peripheral nerves and cranial nerves.

There are two categories of nerve conduction studies: motor and sensory nerve conduction testing.

The basic findings are generally twofold: (1) how fast is the impulse traveling? (e.g., how well is the electrical impulse conducting?); and, (2) what does the electrical representation of the nerve stimulation (action potential morphology) look like on the screen? (e.g., does there appear to be a problem with the shape or height that might suggest an injury to some portion of the nerve such as the axons or the myelin?) [21].

While interpreting NCV, the distance between electrodes and the time it takes for electrical impulses to travel between electrodes are used to calculate the speed of impulse transmission. Slower than normal speed could indicate nerve damage from direct trauma, diabetic or peripheral neuropathy, viral nerve infection or nerve entrapment diseases like the Carpal Tunnel Syndrome among other conditions.

3.4 Neurodiagnostic Instruments

The basic neurodiagnostic instrument include data collection, display and storage. The components of these systems include electrodes, connecting wires, amplifiers, a computer control module including software, and a display and printing device. Evoked potential measurements and nerve conduction studies include an additional stimulators (Fig. 1).

Electroencephalograph, electromyograph/electromyoneurograph and nerve conduction study systems have similar block diagram. The differences are in:

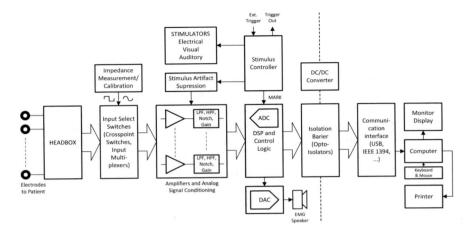

Fig. 1 Generic neurodiagnostic instrument block diagram

sensitivity and number of input channels, characteristic frequencies of low-pass filters, high-pass filters, and optionally notch-filters, types and output characteristics of stimulators, amplitude and temporal resolution of analogue to digital converter, computer performance and application specific software.

Headbox provide an interface between the patient electrodes and input select switches (input multiplexers, crosspoint switches) of the amplifier. Input multiplexers select the proper inputs for recording patient data, conducting an impedance test, or running a calibration procedure. Signal is then amplified by an adjustable gain amplifier to set input sensitivity range from typically 10 μV to 100 mV full scale. Amplified signal is passed through a high-pass filter (HPF) and low-pass filter (LPF). Optionally notch filter reduces amplitude of 50 Hz or 60 Hz line interference. Internally generated calibration pulse is used to check the signal path integrity from the electrode input connectors on the headbox to the host computer.

An impedance measurement circuit lets the user measure the surface electrode impedance and the integrity of the patient-to-electrode connection. Good contacts of the surface electrodes to the skin is one of the important conditions for a good signal. It is necessary to achieve both, low electrode impedance, as well as balanced electrode impedance. Electrode impedance should be below 30 kΩ, but for good recording impedance should be below 5 kΩ and matched within 1.5 kΩ.

The electrically evoked signals are often contaminated by a stimulus artifact. Stimulus artifact suppression implemented in the amplifier may be used to help counter the effects of electrical stimulus on the measured patient signals. This feature may be applied to individual channels or may be enabled or disabled globally.

The analog-to-digital converter (ADC) converts analog signal to a series of digital samples that represent the data. These samples are processed by a digital signal processor (DSP) and transferred to the host PC via communication interface (usually USB or IEEE 1394). Information for synchronizing the data with stimulus events is placed into the data stream as well. Control and status information from the host PC to the system are exchanged along with the data via the communication interface.

EMG signal may be displayed audibly through a speaker. There are two implementations. In the first the filtered analog patient signals are mixed and presented to an audio amplifier and speaker. In the second, the digital patient data samples are mixed and presented via an audio digital to analog converter (DAC) to an audio amplifier and speaker for listening.

Stimulators are controlled by stimulus controller (stimulus pulse generator). This block produces the signal that fires the stimulators. The stimulus pulse from this block goes to each type of stimulator. In addition, external trigger outputs along with a trigger inputs may be available for the control and monitoring of externally generated stimuli. The important role of stimulus controller is synchronization of the stimulus delivery time (MARK) with data collection and processing. If they are present in the system, reflex hammer and footswitch may also be connected to the stimulus controller.

The most commonly used stimulators in neurodiagnostic examinations are electrical, visual and auditory stimulators.

The electrical stimulator is used to provide an electrical signal to the patient, for nerve conduction study (NCS), nerve conduction velocity (NCV), and somatosensory evoked potentials (SEP) recording. There are two types of electrical stimulation: constant current and constant voltage. The constant current amplitude typically range up to 100 mA. The constant current stimulator will try to maintain the selected current level, regardless of voltage required to overcome the skin impedance. If the required amplitude of the voltage is higher than the maximum allowable (maximum compliance voltage), the system displays the warning message. The constant voltage stimulator delivers a selected voltage level to the patient. The constant voltage may range up to few hundred (300–400) volts with the maximum pulse width up to few milliseconds.

Visual stimulators are used for visual evoked potentials (VEP) and electroretinograms recording. The two types of visual stimulators are pattern reversal in which a black and white checker board is generated (on a computer screen), and flashing light. Reversing checkerboard pattern produces a short duration evoked response that is better defined than with flash stimuli. This stimulator is used in clinical environments for recording visual evoked potentials and electroretinograms. Flash stimulators, such as the LED goggles or a strobe light, produce a longer duration response that is less well-defined than checkerboard stimuli, but they can elicit a visual evoked response through closed eyelids. Therefore, LED goggles are primarily used in the operating room or intensive care unit with an anesthetized or otherwise unconscious patient.

The auditory stimulators are used for auditory evoked potentials (AEP) and auditory brainstem response (ABR) recording. There are two basic types of stimulation: air (using ear inserts or headphones), and bone conduction (using bone vibrator). The auditory stimulator provides a click or a tone stimulus.

The three types of click polarity are: rarefaction (the earphone diaphragm moves away from the ear), condensation (the diaphragm moves toward the ear), and alternating (the diaphragm will deliver a rarefaction stimulus and condensation stimulus every other stimulus). The variable click parameters are click rate, duration, intensity level and polarity.

The variable tone parameters are frequency and envelope shape. The envelope variables are rise/fall time and plateau time. Output level of each analog channel is adjustable in 1 dB steps up to 140 dB SPL (sound pressure level). The auditory stimulator also provides noise masking. When stimulus levels higher than 95 dB are delivered to the patient, the stimulus can travel by bone conduction to the non-test (contralateral) ear. Noise masking applied to the non-test ear prevents it from contributing to the evoked response.

The auditory transducers convert the electrical signal from the auditory stimulator to sound. For auditory evoked potential assessment are used headphones (shielded or unshielded), tubal inserts, and bone vibrator. The aural headphones and tubal inserts are used to deliver an auditory stimulus to each ear. The bone vibrator delivers an auditory stimulus through the skull, bypassing the eardrum to stimulate the cochlea.

Stimulators may be internal or external. External stimulators are independent of the computer platform and communicates with stimulus controller module via standard communication interface, usually by the AES (Audio Engineering Society) standard, a high-speed serial interface.

Although the system is fully controlled by a personal computer, some controls and indicators are available on control panel, for example: EMG sound level and Mute switch for the EMG speaker, individual stimulus intensity levels for selected stimulation sites, and stimulator LED (blinking whenever a stim pulse is being delivered).

4 Potential Harms and Hazards Associated with the Use of the Neurodiagnostic Equipment

The main categories of harms and hazards associated with the use of electromedical devices are safety and clinical risks. Safety risks expose patients to potential injury from the equipment. Clinical risks typically relate to misdiagnosis caused by deterioration of characteristics or performance of the medical device over time.

The principles of electrical safety are of great importance in clinical neurophysiology. As with any other electromedical devices with patient applied parts, there is potential risk of electric shock. Furthermore, all of the electrophysiologic studies require the application of electrical connections to equipment that, through connections with the patient, pass small amounts of electrical current to the patient. Although small, there is always an inherent risk to the tissue through which current passes. When more than one medical device is connected to the patient, leakage currents of the devices are summed together.

From the safety point of view, the critical parts of the neurodiagnostic instruments are those in contact with patient. According to base safety standard IEC60601-1 [22] and particular requirements for electroencephalographs (Part 2-26) and electromyographs and evoked response equipment (Part 2-40), "the applied parts of electrical stimulators, visual stimulators, auditory stimulators and biopotential input parts shall be type BF applied parts or type CF applied parts" [23, 24].

Thus, the most important aspect of device safety is to maintain an electrical isolation barrier between a subject connected to an neurodiagnostic device, and the device (typically a computer) to which the neurodiagnostic device is connected. Figure 1 shows simplified block diagram of typical multichannel amplifier with isolation barrier (opto-isolators) in digital signal path. The isolation barrier can be also implemented in input amplifier. In both cases, a medical grade DC-DC converter is used for power supply isolation.

In addition to the issues of electrical safety of all biomedical equipment that relate to the electrical supply voltage and leakage currents, clinical neurophysiology studies such as evoked potentials, nerve conduction studies and transcranial electrical and magnetic stimulation studies, as well as therapeutic devices such as nerve, spinal cord, cortical or deep brain stimulators, involve stimulating neural tissue with electrical currents (or strong magnetic fields), which introduces additional safety considerations related to tissue damage from stimulation and effects on nearby implanted electrical devices such as pacemakers. Besides electrical stimulators, neurodiagnostic equipment may contain visual and audio stimulator whose hazardous output may damage the visual or auditory system. Furthermore, when any part of the neurodiagnostic equipment power supply is interrupted and re-established, all stimulators (electrical, visual, and auditory) shall bi disabled upon power reset, and manual intervention shall be required to re-start any stimulation.

Manufacturers of modern, computer controlled electromedical devices, has taken reasonable measures to ensure that software will remain unaffected by the presence of other, third-party software programs. However, given the vast number of software programs available, manufacturers cannot ensure complete immunity, nor can guarantee immunity against software viruses.

5 Standards and Regulations

The base standard for the medical electrical equipment is given by IEC 60601-1 from the International Electrical Commission, which comprises the General requirements for safety and essential performance. Based on this standard, but with regional deviations and adaptions, are the following standards: United States standard UL-60601-1, Canadian CAN/CSA C22.2 No. 601.1, European standard EN 60601-1, in UK known as BS EN 60601-1, Japanese JIS T 0601-1 and Australian/New Zealand AS/NZ 3200.1.0 [25].

Requirements for the basic safety and essential performance of electromyo-graphs, electroencephalographs and nerve conduction devices are provided by two IEC standards that amend and supplement the general standard IEC 60601-1, 3rd edition [22]: IEC 60601-2-26 Medical electrical equipment—Part 2-26: Particular requirements for the basic safety and essential performance of electroencephalo-graphs, 3rd Ed. [23], and IEC 60601-2-40 Medical electrical equipment—Part 2-40: Particular requirements for the basic safety and essential performance of elec-tromyographs and evoked response equipment, 2nd Ed. [24].

Both particular standards refer to some collateral standards. IEC 60601-2-26 amends the clauses or subclauses of the IEC 60601-1-2 Medical electrical equip-ment—Part 1-2: General requirements for basic safety and essential performance—Collateral Standard: Electromagnetic disturbances—Requirements and tests [26], and the text of this particular standard is additional to the requirements of the IEC 60601-2-27 Medical electrical equipment—Part 2-27: Particular requirements for the basic safety and essential performance of electrocardiographic monitoring equipment [27]. The text of the IEC 60601-2-40 is additional to IEC 60318 (all parts), Electroacoustics—Simulators of human head and ear [28], and ISO 15004-2, Ophthalmic instruments—Fundamental requirements and test methods—Part 2: Light hazard protection [29].

In IEC 60601-2-40 special attention was given to stimulators (electrical, audi-tory, visual): equipment identification, marking and documents, protection against hazardous output, limitation of stimulator output parameters, and power reset behavior of stimulators (automatic re-start of electrical, visual or auditory stimu-lation shall not occur).

6   Inspection of the Equipment

Medical device inspections are aimed to prevent device function deterioration as a potential cause of adverse events and harm to a patient.

The IEC 60601 outlines type-testing in laboratory conditions, but often those conditions are not available or applicable once the device is already in use. To accommodate the needs of in-service device, the standard IEC 62353 Medical electrical equipment—recurrent test and test after repair of medical electrical equipment [30] is introduced. IEC 62353 is specifically designed for testing equipment in the field. IEC 62353 requires testing before initial start-up, after repair, and periodically. The manufacturer is obligated to provide information about the testing of the device, with which the operator of the testing should carefully comply. Inspection must be performed periodically.

According to IEC 62353, the next sequence for electrical safety testing should be followed:

- Visual inspection
- Protective earth resistance

- Leakage current
- Insulation resistance
- Functional test
- Reporting of results
- Evaluation
- Check and prepare for normal use.

Manufacturers of electromedical equipment prescribe in Service Manual detailed system verification procedures, routine backup and maintenance procedures.

As an example, inspection procedure for modern computer-based neurodiagnostic instruments designed to perform electromyography (EMG), nerve conduction studies (NCS), evoked potentials (EP) and electroencephalogram (EEG) should be as follows [31].

System inspection and cleaning

- In desktop computer systems check board seating in the expansion slots and verify the internal cables are securely connected.
- Check for excessive dust accumulation at the power supply and computer vents, and, if necessary, clean/vacuum dust from ventilation slots.
- Visually inspect external system wiring for any damaged or unseated cables, and replace any damaged cables as needed.
- Clean instrument exterior and the system accessories.
- Power on the system and verify that the ventilation fans are turning on.

Computer (Windows operating system)

- Restart the computer and verify that system boots properly.
- Verify that all devices listed in Device Manager are working properly.
- Check the Application, Security and System logs for pattern of warnings or errors.
- Check disk free space and if it is less than the specified, archive and delete the patient exam files on the hard drive. Run Check Disk and check for any error condition.
- Defragment the hard drive.

Built-in diagnostic programs allow comprehensive testing of the system's amplifier and stimulator modules. The diagnostic run in two modes:

- The Auto-Test mode runs without user interaction, and provides a short Pass/Fail report for each tested unit. When the diagnostic software detects a failure, it provides an error code to pinpoint the failure mode.
- The Manual test mode requires user interaction. Separate tests are provided for individual building blocks (for example: headbox and amplifier, control panel, auditory and electrical stimulator).

  - Manual headbox test check signal path integrity from the headbox inputs to the host computer. Headbox integrity test and amplifier calibration test requires defined calibration signal that is routed via the cable to a headbox or amplifier input. Calibration signal is often available on the dedicated base unit's connector. The software compares acquired data with expected value and if the results are within the allowable tolerances, the calibration test passed. If a wave shape-related problem is observed, recommendation is to verify all hardware filters. The generic test for hardware filters is to run a square wave calibration pulse through all amplifier channels, and observe the effects of the filter settings on the shape of the waveform. Test starts with High-pass and Low-pass filter settings wide open. Raising the lower cut-off frequency, the waveform shape gradually changes to a form that resembles the charge/discharge current in a capacitive circuit (Fig. 2). Lowering the higher cut-off frequency, the waveform shape gradually rounds off (Fig. 2).

– Control panel test checks for proper operation of each control panel's potentiometers.
– Auditory test require operator to verify audible click and tone burst signals from the left and right headphones.
– Electrical stimulator test checks electrical stimulus delivery and output level. To check stimulus delivery, depending on stimulator model, operator may stimulate his or her wrist or thumb muscle, or, if applicable, short the + and − probe tips together and compare delivered current intensity with measured. If an electrical stimulator output does not follow the set value, operator may check the stimulator output using a resistive load and oscilloscope. Instrument itself may be also used as a digital oscilloscope by connecting the stimulator output to a load/attenuator test circuit (Fig. 3) and feeding the attenuated signal to system's amplifier. Using the cursors, voltage levels at

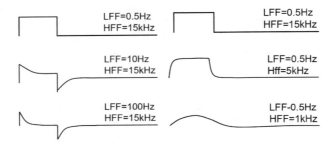

Fig. 2 Effects of lower cut-off frequency (left) and higher cut-off frequency (right) on the shape of the square wave calibration pulse [31]

Fig. 3 Loopback test fixture [31]

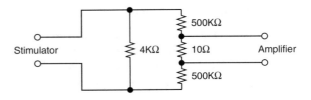

the leading and trailing edges of the waveform (peak and droop points, rise/fall times) should be measured and checked for acceptable tolerances.

Miscellaneous Functions—system may contain various functions and user interfaces required by applications. Each of these units requires a periodical inspection.

- LED goggles

  - Using the visual evoked potential (VEP) test, verify that the left and right LED arrays flash and that all LEDs in each array light up.

- Foot switch

  - Enter the test that uses foot switch (for example, nerve conduction study, NCS) and select recurrent stimulation. Pres the foot switch a few times and verify that the screen message displayed at the screen switches between RUN and STOP. Assure that no contact bounce occurs.

- Reflex hammer

  - Enter the test that uses reflex hammer (for example, nerve conduction study, NCS) and set the stimulator type to Reflex hammer. Tap the reflex hammer against hand. Each tap should trigger a data sweep in the waveform screen.

- Trigger In

  - Enter the nerve conduction study (NCS) test and set the stimulator type to External. According to user manual, provide the appropriate trigger pulse to Trigger In connector. Each trigger pulse should start a data sweep in the waveform screen.

- Trigger Out

  - Trigger out signal is often a standard TTL logic level pulse. Trigger pulse can be measured using an oscilloscope, but if Trigger Out signal is being used to trigger an external stimulator, the easiest way to test the signal is to perform a functional test using the stimulator.

- EMG speaker

  - Enter the measurement of spontaneous EMG activity and set the amplifier sensitivity to 100 µV, to provide an appropriate level of background noise. Turn the Audio volume knob slowly to verify a linear increase in sound from EMG speaker.

- Electrical stimulator current or voltage compliance

  - The output voltage of the current stimulator cannot exceed a particular maximum value determined by construction of the output stage. Manufacturer specifies parameters of the current pulse and the value of the

impedance connected to the stimulator output in order to perform maximum output voltage test.
- Similarly, the output current of the voltage stimulator cannot exceed a particular maximum value. Parameters of the voltage pulse and the value of the impedance connected to the stimulator output in order to perform this test are specified by manufacturer.

Application verification tests include the steps of generating and printing a report, according to system operator or system user's guide. Common applications include, but are not limited to:

- Auditory evoked potential
- Visual evoked potential
- Somatosensory evoked potential
- Motor nerve conduction
- Sensory nerve conduction
- Spontaneous activity EMG.

References

1. Dudek RW, Louis TM (2012) High-yield gross anatomy. Lippincott Williams & Wilkins, Baltimore
2. Guyton AC, Hall JE (2006) Textbook of medical physiology, Eleventh edn. Elsevier Saunders, Philadelphia
3. Mader SS (2011) Human biology, Eleventh edn. McGraw-Hill, New York
4. Penfield W, Rasmunssen T (1950) The cerebral cortex of man: a clinical study of localization of function. MacMillan, New York
5. Noback CR, Strominger NL, Demarest RJ, Ruggiero DA (2005) The human nervous system: structure and function, 6th edn. Humana Press Inc., Totowa
6. Niedermeyer E, Da Silva FL (eds) (2005) Electroencephalography: basic principles, clinical applications, and related fields, 5th edn. Lippincott Williams & Wilkins, Philadelphia
7. Kennett R (2012) Modern electroencephalography. J Neurol 259:783–789
8. Kohden Nihon (2014) Accessories neurology. Nihon Kohden Europe GmBH, Kleinmachnow
9. Medved V (2001) Measurement of human locomotion. CRC Press, Boca Raton
10. Buchtal F, Clemmesen S (1941) On the differentiation of muscle atrophy by electromyography. Acta Psychiat Kbh 16:143
11. Ladegaard J (2002) Story of electromyography equipment. Muscle Nerve 25(S11):S128–S133
12. Stalberg E, Trontelj J (1979) Single fibre electromyography. The Mirvalle Press Limited, Surrey
13. Basmajian JV, De Luca C (1985) Muscles alive: Their functions revealed by electromyography. Williams & Wilkins, Philadelphia
14. Medved V, Cifrek M (2011) Kinesiological electromyography. In: Klika V (ed) Biomechanics in applications. InTech, Rijeka
15. Barbero M, Merletti R, Rainoldi A (2012) Atlas of muscle innervation zones—understanding surface electromyography and its applications. Springer, Italia, Milan
16. Blumenstein B, Eli MB, Tenenbaum G (2002) Brain and body in sport and exercise biofeedback applications in performance enhancement, 1st edn. Wiley, Sussex

17. Eichler W (1937) Ueber di Ableitung der Aktionspotentiale vom menschlichen nerven in situ. Zeit Biol 98:182–214
18. Dawson GD, Scott JW (1949) The recording of nerve action potentials through the skin in man. J Neurol Neurosurg Psychiat 12:259–267
19. Binnie CD, Cooper R, Mauguiere F, Fowler CJ, Prior PF, Osselton JW (1995) Clinical neurophysiology—EMG, nerve conduction and evoked potentials. Butterworth-Heinemann Ltd., Oxford
20. Gilliatt RW, Melville ID, Velate AS, Willison RG (1965) A study of normal nerve action potentials using an averaging technique (barrier grid storage tube). J Neurol Neurosurg Psychiatry 28:191–200
21. Weiss L, Silver J, Weiss J (eds) (2004) Easy EMG. A guide to performing nerve conduction studies and electromyography. Butterworth-Heinemann, Oxford
22. IEC 60601-1:2005 + AMD1:2012 CSV. Medical electrical equipment—Part 1: General requirements for basic safety and essential performance, 3.1 edn. International Electrotechnical Commission, Geneva, Switzerland
23. IEC 60601-2-26:2012. Medical electrical equipment—Part 2-26: Particular requirements for the basic safety and essential performance of electroencephalographs. International Electrotechnical Commission, Geneva, Switzerland
24. IEC 60601-2-40:2016. Medical electrical equipment—Part 2-40: Particular requirements for the basic safety and essential performance of electromyographs and evoked response equipment. International Electrotechnical Commission, Geneva, Switzerland
25. Eisner L, Brown RM, Modi D (2004) National deviations to IEC 60601-1. http://www.mddionline.com/article/national-deviations-iec-60601-1. Accessed 5 Feb 2017
26. IEC 60601-1-2:2014. Medical electrical equipment—Part 1-2: General requirements for basic safety and essential performance—collateral standard: electromagnetic disturbances—requirements and tests. International Electrotechnical Commission, Geneva, Switzerland
27. IEC 60601-2-27:2011. Medical electrical equipment—Part 2-27: Particular requirements for the basic safety and essential performance of electrocardiographic monitoring equipment. International Electrotechnical Commission, Geneva, Switzerland
28. IEC 60318-1:2009. Electroacoustics—Simulators of human head and ear—Part 1: Ear simulator for the measurement of supra-aural and circumaural earphones. International Electrotechnical Commission, Geneva, Switzerland
29. ISO 15004-2:2007. Ophthalmic instruments—fundamental requirements and test methods—Part 2: Light hazard protection. International Organization for Standardization, Geneva, Switzerland
30. IEC 62353:2014. Medical electrical equipment—recurrent test and test after repair of medical electrical equipment, 1.0 edn. International Electrotechnical Commission, Geneva, Switzerland
31. VIASYS Healthcare, Inc. (2005) Neurodiagnostic instruments service manual For Windows XP-based Endeavor CR, VikingQuest and VikingSelect Systems

Part II
Inspection and Testing of Therapeutic Devices

Inspection and Testing of Defibrillators

Milan Ljubotina

Abstract Over the last 50 years of use, defibrillation has been proved to be safe and efficient method to terminate lethal arrhythmias like ventricular fibrillation and ventricular tachycardia without pulse. However, in order to ensure safety and efficacy of this therapy, it is necessary to have a defibrillator which has been tested and proved to be fully functional. Preventive maintenance of all medical devices is highly important to ensure correct diagnosis and therapy, but for defibrillators it is even more important, because a defibrillator is a life-saving device and it is used in the case when a patient life is in danger. Any failure or partial failure of a defibrillator functionality may result in death of a patient. Therefore, there is a zero-tolerance on discrepancies between full functionality of defibrillator, as described in device documentation and actual functionality of defibrillator. Preventive maintenance of defibrillator consists of series of tests which need to be done on the device, with use of special measuring equipment. The most important tests include measurement of output energy which is being delivered from a defibrillator, at all energy levels, to a certain testing impedance. It is also possible to repeat this testing on variable impedance values, and that is highly recommended, because actual patient impedance values will also vary. The documentation for each defibrillator defines tolerance of the delivered energy value, so the actual value must be within prescribed tolerance values, in order to declare the device fully functional. In this chapter, the principles of operation are described too, as well as related standards and the simplest to use kind of defibrillators, called automatic external defibrillators (AEDs).

M. Ljubotina (✉)
University of Zagreb, Zagreb, Croatia
e-mail: milan.ljubotina@kardian.hr

© Springer Nature Singapore Pte Ltd. 2018
A. Badnjević et al. (eds.), *Inspection of Medical Devices*, Series in Biomedical
Engineering, https://doi.org/10.1007/978-981-10-6650-4_8

161

1 Historical Overview of Defibrillators Invention and Development

Ventricular fibrillation (VF) is a heart arrhythmia caused by stimuli coming from multiple places on the heart muscle, resulting in specific effect on the heart ventricles. They are not contracting normally as they should, but only shivering, or—fibrillating. The absence of normal heart contractions causes the blood circulation to stop. The lack of blood supply to heart arteries and to brain causes death in each case, except for the ones when this lethal arrhythmia was stopped.

This lethal arrhythmia can be stopped by delivering of therapeutic dose of electric current to the heart which is in ventricular fibrillation. A current of approximately 2 A, applied directly to the heart muscle through "spoon" electrodes (used during the open-heart surgery) can be enough to stop the fibrillation effectively. The heart will "stop" for a very short period of time, staying in a stage without any electrical activity, an after that phase, a normal rhythm may appear.

In case that the first attempt was not successful, the next one can be repeated, with a higher current. Usually, the current is being increased in three steps, while the fourth and all subsequent trials are being performed with the highest available current.

The process described here is called defibrillation, and device used to deliver the electric shock—defibrillator. The increase of current is performed through selection of different (increasing) energy levels on the defibrillator. Increasing may be done either manually or automatically. Output current which will pass through the patient's heart will depend on different factors, including the patient's thorax impedance, which is highly individual and cannot be predicted. Therefore, a modern defibrillator must be able to adapt to different chest impedance values, in order to deliver optimal current value. The relationship between selected energy and output current will be described later in this chapter.

Ventricular fibrillation is the most common cause of cardiac arrest and sudden cardiac death (SCD). The statistics show that the occurrence of sudden cardiac arrest (SCA) in certain human population per year can be as high as 2‰. However, the probability of successful resuscitation of the patient in VF increases if the first defibrillation shock has been delivered to the patients very quickly, in the first minutes after the occurrence of VF. Therefore, the defibrillator used to resuscitate the patient must be ready and completely functional, because any malfunction or other technical issue may postpone or prevent delivery of the electrical shock to this patient. So, finally, it is the question of saving of human life, and consequently there should be zero tolerance on defibrillator technical issues or malfunctions. That is the reason why preventive maintenance and inspection of defibrillators is vitally important, as the defibrillator is a life-saving device.

History of the trials of scientists and physicians to resuscitate a dying person (or to establish a normal heart rhythm) by use of electricity is actually much longer than it is generally thought. The oldest recorder trial dates from the year 1774. (published in 1778.) Charles Kite, an English physician developed a machine that had a capacitor to store energy, variable output and two electrodes which could be applied anywhere on the body. He applied the machine on a 3-year old girl which fell from a height and looked apparently dead. 20 min after the incident, he started applying electrostatic shocks on different parts of her body, including the chest. After while he noticed that the girl was waking up. However, even Kite suspected that the girl was in coma and not dead, so this was registered only as the first trial to resuscitate the patient with use of electricity.

The real pioneer of cardiac resuscitation with use of the electric shocks is prof. Paul Zoll, MD, (1911–1999) a scientist who worked in Beth Israel Hospital in Boston, MA, USA. He was a cardiologist dedicated to work in three parallel areas at the same time; delivering of medical care to patients, scientific work through original research and teaching of students. His scientific work has created the fundaments to saving of thousands of lives every year around the world, because the defibrillators today are using the discoveries from his scientific work.

In 1950, a presentation at a meeting of the American College of Surgeons, in Boston, about stimulating the sino-atrial node via a transvenous catheter, inspired Zoll to develop a technique for pacing the heart through the intact chest during asystole. With an epochal publication in 1952 he described cardiac resuscitation via electrodes on the bare chest with 2-ms duration pulses of 100–150 V across the chest, at 60 stimuli per minute. This initial clinical description launched widespread evaluation of pacing and the recognition by the medical profession and the public that the asystolic heart could be stimulated to beat; it became the basis for future clinical pacing developments. This technique eventually fell from favour, except in an emergency, because of associated pain and the limited mobility it allowed the patient. It was later revised using larger skin electrodes and longer pulse durations, both of which made the shocks less painful and therefore more acceptable.

In 1955, Zoll described a mechanical technique for "stimulating" the asystolic heart. In 1956, he published a transcutaneous approach to terminate ventricular fibrillation with a much larger shock, of up to 750 V, and later described similar termination of ventricular tachycardia. His use of an alternating current shock began clinical cardioversion–defibrillation but eventually was replaced by direct current shock, largely for technical reasons [1].

From today's perspective, it is almost impossible to imagine that Zoll experienced serious critics to his work after he had published the results of his scientific work. The opponents came from certain circles which were claiming that Zoll is playing God by resuscitating "dead" patients. Meaning that death was an act of God, and no man should stand against it. Fortunately, Zoll was able to justify his scientific work in a way that he discovered a way to save human lives, and that has been recognized and appreciated from that time till now.

From the early years of use of defibrillators, the devices have evolved, as the technology has developed. In Fig. 1, the development of defibrillators is presented through different models, throughout the years.

First portable defibrillator designed by Frank Pantridge in 1965[1]

Defibrillator MRL AMB-PAK, used in 1970's[2]

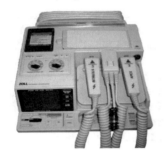

ZOLL PD 1400 defibrillator, 1992[3]

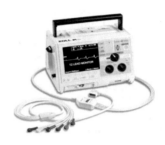

ZOLL M Series defibrillator, 1998[4]

ZOLL E series defibrillator, 2007[5]

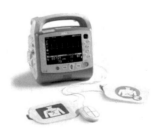

ZOLL X series defibrillator, 2012[6]

Fig. 1 Development of defibrillators through the years. *Source* 1, 2 National EMS Museum http://emsmuseum.org; 3, 4 ZOLL Medical Corporation, www.zoll.com; 5, 6 ZOLL Medical Corporation, www.zoll.com

2 Importance of Preventive Maintenance of Defibrillators

From the previous chapters, it is clear that defibrillators are life-saving devices. There are other medical devices belonging to the same group (risk class), but defibrillators are specific in a way that they are used when the patient is dying. Therefore, a defibrillator must be fully functional in that moment, and there is zero tolerance on any technical issue which may postpone or prevent therapy delivery to the patient. It must work first time, every time. In order to understand the importance of proper preventive maintenance of defibrillators, we will review potential risks involved during defibrillation, for user and patient.

As the defibrillator is delivering a high-voltage electrical shock, it must be secured that the user or bystanders will not be exposed to that shock, or any other electrical current, coming from a device. There are precaution measures which need to be taken during the resuscitation procedure, but at the same time, the defibrillator must be free from any damages, malfunctions or technical issues which may cause that to happen.

American Food and Drug Administration (FDA) has an extensive database with issues caused by technical failures on all medical devices, including defibrillators, and it is publicly available on http://www.accessdata.fda.gov/scripts/cdrh/cfdocs/cfMAUDE/Search.cfm?smc=1. Some of the cases described there have or could have caused health injury ore lethal consequences on a patient treated with such device. Here are some examples of technical issues which could have been avoided with correct preventive maintenance of defibrillators:

- The customer reported that their device would not power on
- Device prompted "connect cable" when used on a patient with cardiac arrest. The users had charged defibrillation energy to administer a shock, but when they pushed the shock button, the device prompted to connect the cable.
- The customer reported that his device would continuously reboot on its own when it was powered on. The device may not be able to be used to deliver defibrillation energy if needed.
- The customer reported that during a patient event while monitoring the patient, the display on their device momentarily went blank, which stopped the monitoring capabilities. When the screen returned, the device appeared to be non-responsive and could not be used to continue monitoring. The customer stated that they were initially performing CPR on the patient, but defibrillation was not needed during the call as they only needed to monitor the patient for the remainder of the call. Post transport, at the arrival of the hospital, the customer indicated that they were still unable to power off the device as it was not responding to their button presses.
- The customer contacted the manufacturer to report that their device did not deliver a defibrillation shock during use on a patient. A 200 J defibrillation shock had been administered to the patient, but the hospital crew wanted to deliver a 360 J shock, the device charged defibrillation energy, but did not deliver the shock. CPR was initiated and the device then reportedly shocked without the shock button being pushed.

It has already been described that defibrillator delivers energy shocks in multiple (most often three) escalating energy levels. Each time, the energy level has been selected on the defibrillator, but the actual delivered energy will vary and in most cases, will not be exactly equal to the selected value. The variation is a result of different factors, including the impedance in the entire electrical circle, where the patient impedance is the most variable part. Each defibrillator manufacturer will define acceptable tolerance of delivered energy on a certain level. For example, it can be ±20% of the selected value.

If the delivered energy is outside that range, it may lead to failure of therapy delivery to a patient. In case that the user fails observe the problem, the consequences may be fatal. Although modern defibrillators will measure and automatically print the values of delivered energy and current, it is essentially important to include the defibrillation simulation in preventive maintenance protocol. It is done by use of defibrillator simulator/tester which will measure and display actual delivered energy on each selected energy level. The testing needs to be repeated on all energy levels available on certain defibrillator. The impedance used is typically 50 Ω, however, certain defibrillator testers will enable the user to change the impedance value, too.

Failure to deliver the shock may come from a depleted battery (in out-of-hospital resuscitation), or a faulty charging circuit which may deliver not enough or zero energy. Further in this chapter, a full protocol for defibrillator preventive maintenance will be explained. Following such protocol in each annual service process minimizes the risk of unwanted issues for user or patient.

3 Principles of Operation

Basic construction of a defibrillator is shown in the Fig. 2.

A high voltage transformer is used to transform the voltage from 220 V to a high voltage of desired level. Further, the voltage generated that way is used to charge a high-voltage capacitor, through the rectifier diode. Instead of single capacitor, an actual capacitor bridge is used. At this point the defibrillator has been charged.

As the "Discharge" buttons on both defibrillation paddles (electrodes attached to patient's bare chest) have been simultaneously pressed, a relay switch which is connected to the capacitor changes its position to connect the capacitor to the second electrical circuit. In that circuit, a capacitor is being discharged through the

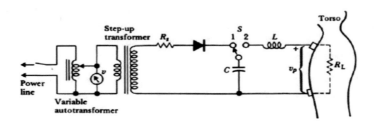

Fig. 2 Electrical diagram of defibrillator. *Source* http://z-diagram.com

patient's chest, over the inductance. This way, the electrical impulse passes through patient's chest and heart muscle and performs defibrillation [2].

A defibrillator can be used also to convert other arrhythmias (like atrial fibrillation), in which the own rhythm of the heart exists, but it is not stable and can develop to a more dangerous arrhythmia, like ventricular fibrillation. In this case, it is necessary to synchronize the defibrillation shock with patient's heart rhythm.

This kind of therapy is called synchronized cardioversion.

In Fig. 3, one QRS complex from patient's ECG is shown. The defibrillator must be able to recognize the R wave, and to mark it with an arrow. The cardioversion shock must be delivered in the interval right behind that point, but prior to the inflexion point between "S" and "T" points on the ECG. The period after that point on the ECG is called vulnerable period. If the cardioversion shock would be delivered during the vulnerable period it would most likely cause a ventricular fibrillation.

In order to avoid that, a defibrillator has a circuit dedicated to enable synchronization for cardioversion. In Fig. 3 it is shown how an ECG amplifier is used to provide input for synchronization circuit. In this case, the synchronization circuit will activate the cardioversion shock delivery in a period after the synchronization marker, before the vulnerable period.

The electric impulse being delivered during the defibrillation has evolved over the years. We will focus here on two most common defibrillation waveforms; monophasic (direct current) and biphasic (alternating current). For many years since their start of use, all defibrillators had monophasic technology. Biphasic defibrillators were first introduced in the late 1980s. At this point, all currently produced defibrillators are biphasic, but there are still monophasic defibrillators being used in healthcare institutions around the world. Here are the descriptions of both waveforms:

Monophasic waveforms: A type of defibrillation waveform where a shock is delivered to the heart from one vector as shown in the Fig. 4. It is shown graphically as current versus time.

In this waveform, there is no ability to adjust for patient impedance, and it is generally recommended that all monophasic defibrillators deliver 360 J of energy in

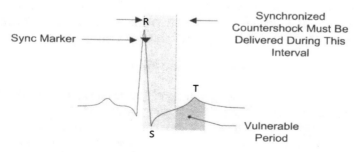

Fig. 3 Synchronization on R-wave. *Source* http:/hackettgroup.org

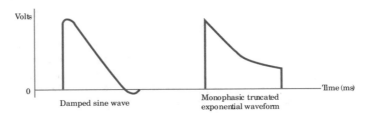

Fig. 4 Monophasic waveform. *Source* British Medical Journal, http://heart.bmj.com

adult patients to insure maximum current is delivered in the face of an inability to detect patient impedance.

Biphasic waveforms: A type of defibrillation waveform where a shock is delivered to the heart via two vectors. Biphasic waveforms were initially developed for use in implantable defibrillators and have since become the standard in external defibrillators.

While all biphasic waveforms have been shown to allow termination of VF at lower current than monophasic defibrillators, there are two types of waveforms used in external defibrillators. These are shown in the Fig. 5.

The upper waveform is called truncated exponential biphasic waveform, while the lower graph shows rectilinear biphasic waveform.

In order to understand the advantages of biphasic technology, we need to clear the relationship among electrical values involved in defibrillation. It has clearly been established that current defibrillates the heart. But, it can be easy overlook the importance of current in defibrillation because defibrillation settings are labelled with energy, not current. Energy is actually the product of three variables:

$$\text{Energy} = \text{Voltage} \times \text{Current} \times \text{Time} \tag{1}$$

The energy setting on a defibrillator corresponds to how much voltage is charged on the capacitor within the device. This correlation is not the same for every device. One manufacturer may charge 1500 V for a 200 J setting, while another may charge 2200 V. For this reason, comparison of energy settings between devices is no longer appropriate.

The amount of current delivered to the heart is a function of two factors: voltage and impedance. The amount of current delivered to the heart is determined by Voltage/Impedance. (This relationship is known as Ohm's Law.)

$$\text{Current} = \text{Voltage}/\text{Impedance} \tag{2}$$

So, for a given energy setting, the current depends on (1) the amount of voltage used for a given energy setting on a particular device, and (2) patient impedance.

It is also important to distinguish between two different types of current: average current and peak current. Peak current is the maximum amount of current seen by the heart. Too much peak current can result in electroporation (formation of

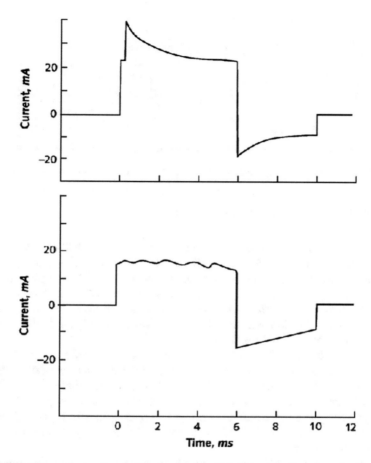

Fig. 5 Biphasic waveforms. *Source* Annals of Internal Medicine http://annals.org/

aqueous pores in the cell membrane caused by a strong external electric field; the basic mechanism of tissue injury on high-voltage electric shock) of the myocardial cells, which may result in myocardial dysfunction. Average current is the mean amount of current seen by the heart over the duration of the shock. This is the component believed to be responsible for successful defibrillation (Fig. 6).

The goal of a defibrillation shock is to deliver the appropriate amount of average current while minimizing peak current. This is the reason for differences between certain biphasic waveforms, as some of them have developed over time as a result of practical experiences and scientific research.

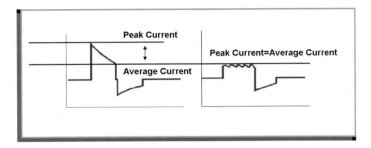

Fig. 6 Peak current and average current. *Source* ZOLL Medical Corporation, www.zoll.com

3.1 Automatic External Defibrillators (AED)

The scientific research has shown that the time from SCA to defibrillation is crucial in a way that the first defibrillation shock must be delivered to the patient as soon as possible. The optimal time frame is no longer than 3–5 min. But, the chances of survival are higher as the time between SCA and defibrillation is shorter. After 5 min, the victim loses 10% of chances for survival with every minute that passes. So, the obvious question was how to assure availability of a defibrillator everywhere in the shortest time possible.

The answer to that question is an Automatic External Defibrillator (AED). An **AED** is an easy to use, safe and reliable defibrillator that automatically diagnoses ventricular fibrillation and ventricular tachycardia. In case that either of those 2 rhythms is detected, the AED will automatically charge to an appropriate energy level, and will give a voice and text prompt to a rescuer to deliver the defibrillation shock. Alternatively, so called fully automatic defibrillators will deliver the chock without user intervention.

The most advanced AEDs have the CPR feedback functionality incorporated. It means that the AED will guide the rescuer (through voice and text prompts) how to perform chest compressions (CPR) correctly. That function is important and helpful in order to maximize chances of survival for a victim being resuscitated by layperson, prior to arrival of the EMS team.

Due to their simplicity of design and use (only 2–3 buttons), AEDs are designed to be simple to use for the layperson. Therefore, the AEDs are today widely used and deployed in places where the large groups of people gather, and therefore, the risk of SCA is higher. The most common places for AEDs installation are; public squares, airports, bus and train stations, hotels, swimming pools, beaches, sport halls, office buildings and big companies, etc.

In Fig. 7, 3 different AED models are shown.

There are several reasons why an AED may play a crucial role in resuscitation of an SCA victim; it gives a unique opportunity to a lay rescuer to give a first aid, and to use the defibrillator in a safe and effective way. Without an AED, the first defibrillation shock may be given only when the EMS team comes to the scene.

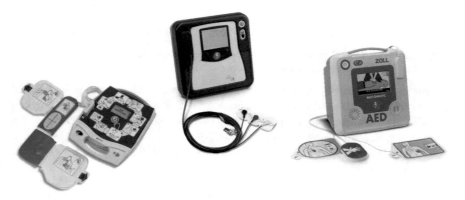

Fig. 7 Automatic External Defibrillators (AEDs). *Source* ZOLL Medical Corporation, www.zoll. com

In many cases the EMS cannot come the scene in 3–5 min. So, the use of the AED during the first aid (CPR) performed by the lay rescuer is a "zero step" in resuscitation process, which may be crucial to enable the professional EMS team to proceed with the following steps in resuscitation, and finally, to successfully save the life of a patient.

However, it is important to understand that a lay rescuer, in many cases, has the opportunity to use the defibrillator on the victim, maybe once in his lifetime. Also, an AED can be installed in some public place and not used in resuscitation for many years. But, in that one case, when the resuscitation really happens, the rescuer should do his/her best, and the defibrillator must be fully functional, without any technical issues.

So, the AEDs, as well as professional defibrillators require proper service and preventive maintenance in order to assure full functionality.

4 IEC 60601—Basic Standard for Medical Devices— Requirements for Defibrillator Design and Testing

The IEC 60601 standard contains requirements regarding the electrical safety of both mechanical and electrical issues in medical devices. It contains 2 parts; IEC 60601-1 and IEC 60601-2. Each part contains further basic and collateral standards [3].

IEC 60601-1-X (X is a number between 1 and 11) is the primary standard and has (sub) standards directly relating to the safety of medical equipment.
IEC 60601-2-X (X is a number between 1 and 58) represents the set of standards is specific to various types of medical equipment and provides additional information to the collateral standards.

The requirements for defibrillators are described in the following IEC standard:

IEC 60601-2-4 medical electrical equipment part 2: particular requirements for the safety of cardiac defibrillators and cardiac defibrillators monitors

This standard actually represents a set of requirements which need to be fulfilled in order to assure safe and reliable functionality of defibrillator. In the next chapter, we will give an example of a regular preventive maintenance protocol for a defibrillator, in accordance with IEC 60601-2-4.

IEC 60601 has also defined certain names and definitions, as well as symbols and markings used in documentation.

The preventive maintenance process is oriented to find any technical issues, failures or discrepancies which may affect defibrillator functionality. In order to do that, several measurements need to be done on the defibrillator, using appropriate tools and testing equipment, in order to measure the **output values** on the device and to compare them with limits defined by the manufacturer. On a defibrillator, it will include safety testing (leakage currents) with electrical safety tester (earth, enclosure and patient leakage currents), as well as measurement of output defibrillator energies on all energy levels. Prior to that part a visual inspection needs to be done, and all steps need to be appropriately documented and all measured values recorded.

5   Regular Preventive Maintenance of Defibrillators

Importance of regular maintenance tests on defibrillators has been clearly described in the previous chapters. In this chapter, we will describe standard steps in a regular maintenance test.

A modern defibrillator is actually a multi-functional defibrillator-monitor unit with several optional modules. Most of the options come from the monitoring side of device. During the preventive maintenance process, all modules and functions need to be tested. Here is the list of modules which can be included in a standard defibrillator-monitor:

- Defibrillator (asynchronous/synchronous)
- External transcutaneous pacemaker (asynchronous/synchronous)
- ECG monitor (3/5 lead or 12-lead)
- SpO2 monitor (pulse oximeter)
- NIBP monitor (non-invasive blood pressure)
- CO2 monitor (carbon dioxide)
- IBP monitor (invasive blood pressure)
- Temperature monitor
- Memory module (data storage)
- Data communications module (serial, Bluetooth, Wi-Fi…)

In order to assure full functionality of a defibrillator all the time, it is common to define for each defibrillator two checkout procedures: the operator's shift checklist and the extensive 6-month or 12-month maintenance tests checkout procedures. Because the defibrillator must be maintained ready for immediate use, it is important for users to conduct the Operator's Shift Checklist procedure at the beginning of every shift. This procedure can be completed in a few minutes and requires no additional test equipment. Modern defibrillators have built-in self-test module which enables the operator to simulate a defibrillation on each energy level. This way, the functionality of almost all parts which are vital for sock delivery are being checked.

A qualified biomedical technician must perform a more thorough maintenance test checkout every six or twelve months to ensure that the functions of the unit work properly. This chapter describes the step by step procedures for performing such maintenance test checkout.

Preventive maintenance includes the following tests [4]:

- Physical Inspection of the Unit
- Front Panel Buttons
- 3, 5, and 12 Leads ECG
- Leakage Current
- Paddles
- Synchronized Cardioversion
- Shock
- Pacer (External pacemaker)
- SpO2 Monitor
- EtCO2 Monitor
- NIBP Monitor
- IBP Monitor
- Temp Monitor

Before start of preventive maintenance testing, following equipment should be prepared:

- Defibrillator Analyzer
- Safety Analyzer
- ECG Simulator; 12 Lead Simulator for 12 Lead test
- An adapter may be required to connect the device to defibrillator analyzer
- Paddles (if used)
- Printer Paper
- Battery
- AC line cord

Further in this chapter, we will give a description of preventive maintenance testing procedure for one defibrillator-monitor. Exact procedure will vary for different manufacturers and models, but the basic steps are the same. All steps described are intended to check in details each module on the unit, to assure full functionality.

Each performed test needs to be verified in the previously prepared report (document).

Note: The protocol below represents an example of preventive maintenance test of important defibrillator modules. For testing of certain defibrillator model, please refer to original manufacturer's technical documentation, and respect exact values for measured parameters, as well as specific requirements which are result of configuration or construction.

5.1 Physical Inspection of the Unit

Housing

- Is the unit clean and undamaged?
- Does the unit show signs of excessive wear?
- Does the handle work properly?
- Does the recorder door open and close properly?
- Are input connectors clean and undamaged?
- Are there any cracks in the housing?
- Do the front panel, buttons or switches have any damage or cracks?
- Are there any loose housing parts?

Cables

- Are all cables free of cracks, cuts, exposed or broken wires?
- Are all cable bend/strain reliefs undamaged and free of excessive cable wear?

Paddles

- Do the adult and pedi plates have major scratches or show signs of damage?
- Do the adult shoes slide on and off easily to expose the covered pedi plates? (if the paddles are constructed this way)
- Are the paddles clean (e.g., free of gel) and undamaged?

Battery (some models have more than one battery)

- Place battery in battery well.
- Is the battery seated in the battery well correctly?

5.2 Front Panel Button Test

Tools Needed: Defibrillator Analyzer.
Test Setup: Install strip chart paper into the recorder tray (some defibrillators may not have the chart recorder).

- Install the battery in the unit or connect the A/C power cord to the unit and then plug the cord into an electrical outlet.
- Connect the defibrillation cable and ECG cable (3 lead, 5 lead, or 12 lead) to the simulator, and defibrillator analyzer
- Turn the device ON
- Confirm that the monitor is on.
- Set the analyzer to NSR of 120 BPM. To check the size of the ECG waveform, press the **SIZE** button.
- As you press the SIZE button multiple (e.g. 0.5, 1.0, 1.5, 2.0, 3.0), the size of the ECG waveform appropriately changes on the display.
- Press the **ALARM ON/OFF** button (**ALARM SUSPEND** on some models). Alarm symbol changes from disabled to enabled.
- Press the **RECORDER** button. The strip chart paper moves out of the unit from the paper tray. Check that the
- Correct time, date, ECG lead annotation and waveform are recorded on the paper. (Set Time and Date, if necessary.)
- Open the paper compartment door, then press the **RECORDER** button.
- The *CHECK RECORDER* message appears on the monitor.
- Close the paper compartment door, then press the **RECORDER** button.
- The strip chart paper flows out of paper tray and the monitor no longer displays the *CHECK RECORDER* message.
- Press the **RECORDER** button. The strip chart paper stops flowing out of the paper tray.
- Move the selector switch to DEFIB. (If such selector exists.)
- Press the **CHARGE** button. The display shows that the unit is charging. The **SHOCK** button lights when the
- unit is charged, and the Ready tone for DEFIB sounds.
- Press and hold the **ENERGY SELECT** buttons (up and down), or rotate the selection knob, until you reach the maximal energy setting.
- Press the **CHARGE** button. The display shows the unit charged and the **SHOCK** button lights.
- Press the **SHOCK** button. The unit discharges and the **SHOCK** button is no longer illuminated.

5.3 3, 5, and 12 ECG Leads Test

Tools Needed: 3 lead, 5 lead, and 12 lead cables. Test each cable separately. Defibrillator Analyzer.

- Connect the lead wires appropriate to each test to the Defibrillator Analyzer.
- Turn on the **MONITOR**.

- Disconnect one lead from the simulator. The monitor displays the *ECG LEAD OFF* message (if configured).
- Reconnect the lead. Repeat step 3.2 with the remaining leads.
- The *ECG LEAD OFF* message appears when the lead is disconnected and clears the lead is reconnected.
- Repeat he steps above 3.2 and 3.3 for 5 lead and 12 lead cables.

5.4   Leakage Current Test

Tools Needed: See the manufacturer's instructions or supplied specifications for the leakage tester you use.
Test Setup: See the manufacturer's instructions or supplied specifications for the leakage tester you use. Repeat leakage test with accessories: defibrillation cable, external paddles, and anterior/posterior paddles.

   Maximum Leakage Acceptance Limits—refer to defibrillator's manufacturer instructions for exact values.

5.5   Paddles Test (If Applicable)

Tools Needed: Defibrillator Analyzer.
Test Setup: If applicable, connect the cable to the paddles and place the paddles in paddle wells.

- Turn the defibrillator ON. Press and hold the **ENERGY DOWN** button on the paddle, or select the lowest energy level on the rotating knob.
- Press the **CHARGE** button.
- Press and release the **APEX SHOCK** button. The unit does not discharge.
- Press and release the **STERNUM SHOCK** button. The unit does not discharge.
- Press and hold both paddles **SHOCK** buttons. The unit discharges.
- Repeat the steps above for each energy level available one the defibrillator.

5.6   Heart Rate Display Test

Tools Needed: ECG Cable (3 or 5 leads), Defibrillator Analyzer.

   Test Setup:

- Turn the selector switch to MONITOR. Press **LEAD** button until "I" displays.
- Connect the ECG leads to the QED 6 Defibrillator Analyzer.

- Connect the ECG cable to the unit.
- Set the Defibrillator Analyzer to 120 BPM.
- Verify that the Heart Rate displays as 120 ± 2 BPM.

5.7 Synchronized Cardioversion Test

Tools Needed: Defibrillator Analyzer.
Test Setup: Connect the defibrillation cable to the defibrillator analyzer.

Set ECG on analyzer to 60–120 BPM.
Note: Sync markers display on the monitor. The sync marker appears as a down arrow over the ECG R-wave peaks n strip chart and display.

- Press **LEAD** button to select **PADS** and Size x1.
- Press the **SYNC** softkey on the defibrillator.
- Enter synchronized cardioversion timing test mode on the defibrillator analyzer.
- Sync markers appear on display.
- Select maximal energy, and then press the **CHARGE** button. When the **SHOCK** button lights, press and hold the **SHOCK** button.
- On the analyzer display, the R-wave to shock delay is less than 60 ms.

5.8 Shock Test

Tools Needed: Defibrillator Analyzer.
Test Setup: Connect the defibrillation cable to the defibrillator analyzer.

- Ensure that a fully charged battery is installed in the unit.
- Turn the defibrillator ON.
- Press the **ENERGY SELECT** down arrow till the lowest energy level. (Or rotate the knob down.)
- Press the **CHARGE** button. Wait for the **SHOCK** button to illuminate.
- Press the **SHOCK** button. The unit discharges into the simulator.
- Read delivered energy from the analyzer and record it.
- Increase the energy level for one step.
- Repeat the steps above for each energy level available one the defibrillator.
- Typically, the resistance in defibrillator analyzer used for this test is 50 Ω. If the defibrillator analyzer enables impedance change, than each step should be performed on different impedances, in order to simulate different kinds of patients.

5.9   Pacer Test

Tools Needed: Defibrillator Analyzer with pacer option included.

For an external non-invasive pacer analyzer, follow the manufacturer's guidelines for measuring the frequency (ppm), output (mA) and the pulse width measured in milliseconds. Note that the analyzer pace load resistor must be less than 250 Ω.

Test Setup:

- Connect the cable from the defibrillator to Defibrillator Analyzer.
- Turn the unit to Pacer mode.
- Set the **PACER OUTPUT** to 14 mA and disconnect cable connector from the analyzer.
- The unit displays the *CHECK PADS* and *POOR PAD CONTACT* messages and the pace alarm is active.
- Reconnect the cable to the analyzer.
- Set rate to 180 ppm; output to 0 mA. Output appears on the analyzer.
- Increase the output to 120 mA. Output on the analyzer is 120 ± 6 mA.
- Increase the output to 140 mA. Output on the analyzer is 140 ± 7 mA.
- Decrease the output to 40 mA. Output on the analyzer is 40 mA or ± 5 mA.
- Increase the output to 60 mA.
- Decrease the rate to 30 ppm.
- Pacer rate on analyzer is 29–31 ppm.
- Increase the rate to 180 ppm. Pacer rate on analyzer is 177–183 ppm.
- Connect the ECG cable to the defibrillator and analyzer. Select the ECG at 60 BPM on the analyzer. Decrease the pacer rate on the unit to 58 ppm.
- The unit displays ECG at 60 BPM with no stimulus markers.
- Press the **Async** Pace softkey. The unit displays ECG at 60 BPM with stimulus markers, and displays the *Async pace* message.
- Turn off the analyzer. Set Pacer Rate to 100 ppm. Press the **RECORDER ON** button.
- The unit displays pace stimulus markers every 15 ± 1 mm.

For preventive maintenance test procedures on monitoring options; SpO2, NIBP, EtCO2, IBP and Temperature, please refer to chapter "Monitors" in this book.

6   Final Remarks

From all arguments written before, in this chapter, it is clearly visible that defibrillators are life-saving devices which have zero tolerance on technical issues. A failure of a defibrillator during the resuscitation may cause a fatal consequences for a patient.

The issue is especially sensitive due to the fact that the use of defibrillator is random and cannot be planned. Besides, it can easily happen that certain defibrillator (especially an AED in public use) is not used for a long period of time. But, when there is a need for a defibrillator, than it needs to be instantly fully functional, without any technical issues.

It is possible to minimize risk of an incident with unwanted consequences if users and technical service personnel follow the manufacturer's rules for daily checks and 6-months or 12-months preventive maintenance testing. The procedure described before gives an overview of main steps in preventive maintenance testing. If the manufacturer's recommendations given in the service documentation are strictly followed, it will enable successful use and lifesaving with a defibrillator over a number of years.

References

1. Cohen S (2015) Paul Zoll MD—The pioneer whose discoveries prevent sudden death. Free People Publishing, Salem
2. Šantić A (1995) Biomedicinska elektronika. Školska knjiga, Zagreb, pp 225–230
3. Backes J (2007) A practical guide to IEC 60601-1. Rigel Medical, Peterlee, pp 2–3
4. ZOLL E Series® Service Manual Rev. D (2014) ZOLL Medical Corporation, Chelmsford

Inspection and Testing of Respirators and Anaesthesia Machines

Baki Karaböce

Abstract Respirators are used in intensive care units and in operating rooms. It consists of filtering, air compression, and humidifying control board units. A respirator is a device that combines the patient's respiratory tract to assist the respiratory system in conditions where the patient has difficulty in breathing or after operations. The device supplies controlled air to the patient by the inner compressor. The breakdown of the oxygen sensor and the heating of the circuit boards (if the filter is not cleaned) are the most common problems in respirators. They may not stabilize with required values over time and the tester is used to maintain stability. The device must be calibrated regularly or if the gauge of the test device does not match the standard values of gas flow, volume, pressure and oxygen parameters. The anaesthesia machine delivers pressurised medical gases like air, oxygen, nitrous oxide, heliox etc. and controls the gas flow individually. It composes a known and controlled gas mixture at a known flow rate and then delivers it to the gas outlet of the machine. Therefore, the fresh gas flow is serviced to the anaesthesia circle breathing system in order to make artificial respiration in the patient and monitor vital functions closely. For patient safety, the most important thing is to check out the system regularly and in pre-use and to ensure that there exists a ready and functioning alternative solution for ventilating the patient's lungs.

1 Introduction

Respirators are an important topic in the health field. The majority of employees have exposure to hazardous gases, vapours, dusts, or mists that require or even suggest the use of a respirator. Some employees may be benefited by the use of a particulate mask while doing certain temporary dusty tasks.

B. Karaböce (✉)
TÜBİTAK UME, National Metrology Institute of Turkey,
Kocaeli, Turkey
e-mail: baki.karaboce@tubitak.gov.tr

© Springer Nature Singapore Pte Ltd. 2018
A. Badnjević et al. (eds.), *Inspection of Medical Devices*, Series in Biomedical
Engineering, https://doi.org/10.1007/978-981-10-6650-4_9

A medical ventilator is a mechanical blower system that is designed to transport breathable air into the lungs and then air out of the lungs in order to supply breath for patients who are unable to breathe or insufficiently breathe physically. Wide ranged and certain types of ventilators cover modern ventilators that are computerized machines and simple manually operated bag valve masks. Ventilators are mainly utilized in anaesthesia machines, in intensive care medicine, emergency medicine and home care.

Medical ventilators are also called "respirators" which may not represent them correctly. So, ventilators and respirators are different functions of medical devices.

The use of mechanical ventilation starts with the various versions of the iron lung which is a type of non-invasive negative pressure ventilator. The iron lung was broadly used during the infantile paralysis epidemics in the 20th century. The following developments were presented by John H. Emerson in 1931 and the Both respirator in 1937, after the promotion of the "Drinker respirator" in 1928 [1].

There are other types of non-invasive ventilators that are also used extensively for infantile paralysis epidemics patients. These are:

- The rocking bed
- Biphasic Cuirass Ventilation
- Positive pressure machines (somewhat simple).

In 1949, John H. Emerson developed a mechanical aid for anaesthesia with the support of the anaesthesia department at Harvard University. Then mechanical ventilators started to be used widely in anaesthesia and intensive care during the 1950s. Their development was stimulated both by the increasing use of muscle relaxants during anaesthesia and the need to treat infantile paralysis patients. Relaxant drugs paralyze the patient and improve operating conditions during surgery, but also paralyze the respiratory muscles.

The East Radcliffe and Beaver models were early examples as can be seen in Fig. 1 in the United Kingdom [2, 3].

A respirator is designed to protect the mask/face piece, hood or helmet that is utilized to protect the patient/user against a various kinds of harmful airborne agents. OSHA's respirator standard, 29 CFR 1910.134, requests the use of respirators to protect workers from breathing contaminated and/or oxygen-deficient air if efficient engineering techniques and arrangements are not applicable, or while they are being established [4]. Some of the other OSHA regulations also require the use of respirators. There is a significant difference between OSHA requirements with regard to particulate masks and respirators.

Respirators must be chosen on the basis of risks to which the worker is subjected too (i.e., particulates, vapours, oxygen-deficiency, or a combination). OSHA also asks for the use of certified respirators. The National Institute for Occupational Safety and Health (NIOSH) certifies respirators.

The chronological evolution of ventilators is summarized below:

- In 1952, Roger Manley produced a ventilator which was fully gas driven in Westminster Hospital, London. It was an optimal design and became the most preferred device by European anaesthetists for four decades. It has no independency for electrical power, and induces no explosion hazard.

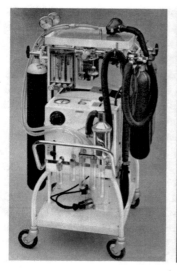 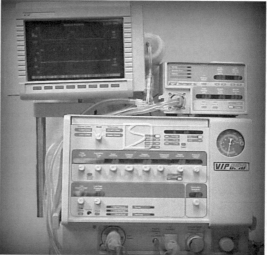

Fig. 1 From the 20th century, an East-Radcliffe respirator model (left) and the Bird VIP Infant ventilator (right). *Source* https://commons.wikimedia.com

- In 1955, the "Bird Universal Medical Respirator" was released by Forrest Bird in the United States of America. Mechanical ventilation was realized when a small green box became a well known part of medical devices. The unit was presented as the Bird Mark 7 Respirator. It was a pneumatic device that does not require an electrical power source for operation.
- In 1971, the Elema-Schönander company released the first SERVO 900 ventilator. It was a revolutionary device around the world for intensive care environments. It was a small, low noise and effective electromechanic ventilator. This device could supply adjusted volume for the first time.
- In 1979, the Model 500A ventilator was introduced by Sechrist Industries. It was specifically produced to use with hyperbaric rooms.
- In 1991, the SERVO 300 ventilator model was presented. The SERVO 300 series supplied a fully new and unique gas delivery system design with a fast flow-triggering response. The platform of this series enabled to treat all patient categories from neonate to adult.
- In 1999, a compact and a smaller LTV (Laptop Ventilator) model were presented into the medical market. This new design opened up an opportunity of mobility for patients with the same functionality.
- In 2001, a modular concept was introduced with the SERVO-i. It gives an advantageous that the hospital has one model of a ventilator for different user needs. It is possible to select the options/modes, software and hardware required for a particular patient category with this new modular concept.

An anaesthesia machine that delivers gases and inhalation agents has a facility for patient monitoring as well as ventilation and safety features. Safety features of the anaesthesia machine have been adopted step by step within years [5, 6].

In a survey conducted between 1962 and 1991 by ASA, 72 of 3791 malpractice lawsuits were founded to be related to gas delivery equipment within an anaesthesia machine. Death and permanent brain damage have been reported as 76% of all the claims. Improper usage of equipment and use without calibration and test were determined as 3 times more than the common failure of equipments in this survey. Surveys indicate the necessity of regular calibration and test of devices for safety use of anaesthesia machines.

General anaesthesia was presented firstly in 1846 by WTG Morton at the Massachusetts General Hospital. Prominent improvements in methods, devices and drugs have made anaesthesia safe over the years. A British anaesthetist H.E.G Boyle, developed a new continuous flow anaesthesia machine in 1917. This anaesthesia machine was eventually patented by the British Oxygen Company as "Boyle's Machine". Several improvements in the simple machine made it easier and safer to control anaesthesia, compared to earlier methods. After significant developments of the Boyle's machine by means of convenience, functionality, mobility and safety, it's being replaced by the "anaesthesia delivery unit" which was also called the "Anaesthesia Workstation" since the 1990s as can be seen in Fig. 2.

An anaesthesia machine that delivers gases and inhalation agents has a facility for patient monitoring as well as ventilation and safety features. Safety features of anaesthesia machines have been adopted step by step within approximately a 100 years period from 1917. Improvements have been made after each problem or accident during application for medical purposes. The developments in anaesthesia machines and systems never stopped within the years by understanding the specifications and features as a point of safety standards every time.

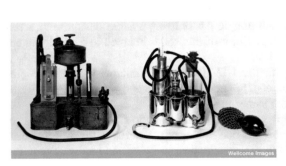 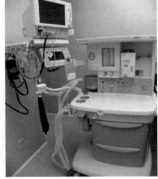

Fig. 2 Old anaesthesia machines from 1920 (left), modern anaesthesia machine (right) *Source* https://wellcomeimages.org (left)/and https://commons.wikimedia.com (right)

Activities of the anaesthesia may create the risk of complications for the patients. The risks can be the operations of the surgeon and/or the collapse or malfunction of the anaesthesia device [7–10]. In the 1990s from an American report, most of the complications of anaesthesia devices were outlined as 23% death, 21% nerve injury, 9% brain damage etc. [11].

If FDA, MAUDE—Manufacturer and User Facility Device Experience database is searched with the following keywords "ventilator, continuous, facility use" as product class and "death" as event type, 163 events can be found in 2016 [12]. Those events may arise from the device and/or user.

2 The Principle for the Work of Respirators and Anaesthesia Machines

The principle of operation can be outlined as [13]:

- an incoming gas flow lifts a weighted bellows unit,
- unit falls intermittently under gravity, and
- it forces to breathe gases into the patient's lungs.

The inflation pressure can be changed by sliding the movable mass on top of the bellows. The volume of the gas supplied is adjustable using a curved slider, which restricts the bellows tour. Residual pressure after the accomplishment of expiration is also configurable by using a small weighted arm that can be visualized on the front panel. This is a robust part and its availability encouraged the introduction of positive pressure ventilation methods into mainstream anaesthetic practice.

A modern positive pressure ventilator mainly consists of:

- a compressible air tank or turbine,
- an air and oxygen supply units,
- valves and tubes set, and
- a reusable and disposable "patient circuit".

The air tank is pneumatically compressed a few times a minute to supply air in the room, or in most conditions, an air-oxygen mixture to the patient. If a turbine is used, the turbine moves air through the ventilator, with a flow valve levelling pressure to provide patient-specific parameters. When excess pressure is released, the patient will exhale passively due to the lungs' elasticity, the exhaled air being released usually through a one way valve within the patient circuit called the patient manifold.

Ventilators may also be furnished with display and alarm systems for patient-related parameters (e.g. volume, pressure, and flow) and ventilator function (e.g. power failure, mechanical failure, and air leakage), backup batteries, oxygen reservoirs, and remote control. The pneumatic system is often replaced by a computer-controlled turbo pump nowadays.

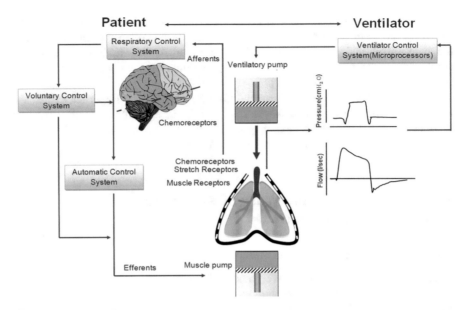

Fig. 3 Modern ventilation system

Modern ventilators are automatically controlled by a small embedded system to allow precise adaptation of flow and pressure characteristics to an individual patient's requirements as seen in Fig. 3. Fine-tuned ventilator adjustments also serve to make ventilation more tolerable and comfortable for the patient [14, 15]. Respiratory therapists are responsible for tuning these settings while biomedical technologists are responsible for the maintenance in the United States and Canada.

The patient circuit is generally composed of a set of three durable, yet light-weight plastic tubes, separated by function (e.g. inhaled air, exhaled air, and patient pressure). Determined by the type of ventilation required, the patient-end of the circuit may be either invasive or non-invasive.

Non-invasive techniques, which are satisfactory for patients who require a ventilator only while resting and sleeping, mainly employ a nasal mask. Invasive techniques need intubation, which for long-term ventilator dependence will normally be a tracheotomy cannula, as this is much more practical and comfortable for long-term care than larynx or nasal intubation.

A basic anaesthesia machine consists of three fundamental systems:

- Gas supply and control
- Breathing and ventilation
- Scavenging.

Commonly an anaesthesia machine is the continuous flow rebreathing through an anaesthesia device. The exhaled gas from the patient (breath) is supplied back to the

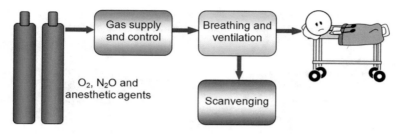

Fig. 4 Block diagram of a basic anaesthesia device

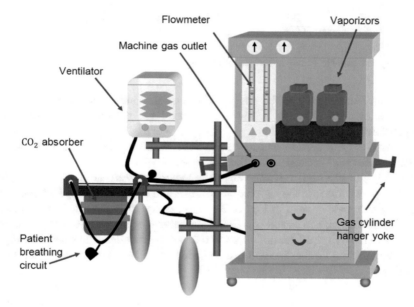

Fig. 5 Basic anaesthesia device

patient after it is processed and mixed with a ratio of fresh anaesthetic gases as seen in Fig. 4.

Normally, oxygen and nitrous oxide gases are supplied from the main system of the wall outlets as 345 kPa pressure. Oxygen flows through a check valve (one direction valve) which is reduced to about 110 kPa by the second stage oxygen regulator before it reaches the flow control valve of the oxygen flowmeter as seen in Fig. 5.

Nitrous oxide gas from the wall outlet passes through the pressure sensing shutoff valve and reaches the nitrous oxide flow control valve of the nitrous oxide flowmeter. The shut off valve is kept open by the oxygen pressure that is normally at 345 kPa. If the oxygen pressure drops to below 172 kPa, the valve will shut off

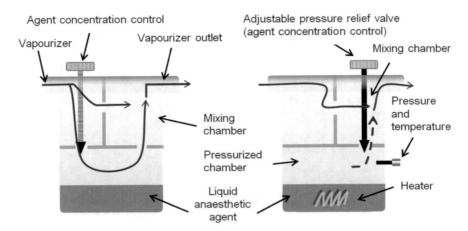

Fig. 6 Bypass variable vaporizer and electronic vaporizer

the nitrous oxide supply to the device. These mechanisms will protect the patient from unknown breathing in a low oxygen level gas mixture in case of supply oxygen failure.

Anaesthesia can be adjusted to a suitable mix and flow of oxygen and nitrous oxide gas mixtures by regulating the flow control valves.

The O_2 and N_2O gas mixtures enter the vaporizer from the inlet and split into two flow paths, one into the vaporizing chamber and the other through a bypass into a mixing chamber as seen in Fig. 6. The gas mixture flowing into the vaporizing chamber flows over a reservoir of a liquid anaesthetic agent. Then the gas meets and mixes with the bypassed gas and flows to the vaporizer outlet. The anaesthetic agent is pressurized into liquid state and heated inside the agent chamber in the electronic vaporizer.

The function of breathing and the ventilation subsystem of an anaesthesia system is to deliver the anaesthetic agent gas mixture to the patient. Most anaesthesia machines deliver a continuous flow of anaesthetic gas and oxygen mixture to the patient. Figure 7 shows the circle of the breathing/ventilation subsystem of an anaesthesia machine under ventilator mode. The scavenging subsystem for waste anaesthetic gas removal is also shown in the diagram.

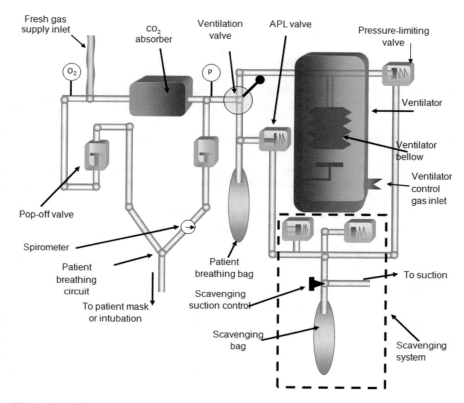

Fig. 7 Breathing/ventilation and scavenging subsystems

3 Safety and Performance Tests

OSHA regulations obligate that employers who are exposed to harmful levels of hazardous gases or vapours must use respirators and apply protocols and procedures. The following work-site for specific items needs to be addressed for a respirator application program.

- Medical evaluations for all workers in need of use of respirators
- Procedures for choosing respirators for different locations
- Fit-testing procedures for tight-fitting respirators
- Procedures for convenient respirator use during routine check and potential emergency cases
- Procedures for cleaning, disinfecting and inspection of respirators.

Normally, a full operational verification procedure maintained by the facility and usually based on the manufacturer's recommendations must be realized periodically. These procedures should be described in a facility's policies and procedures.

The quick tests that should be handled and reported in order to satisfy the safe use of a ventilator are listed below:

- The machine's battery backup and its disconnection alarms should function properly.
- Test lamps must be functioning according to the manufacturer's procedures.
- Appropriate activation of all audible and visual alarms must be tested by using a test lung.
- Proximal airway pressure gauge and positive end expiratory pressure must be controlled.
- Leak tests must be performed either as the machine allows.
- The manometer should be set to the maximum level and the high pressure alarm should activate.
- Plateau pressure should be observed when the ventilator cycles.
- Set the mode to be used for the patient. Verify the proper operation for that mode as the ventilator cycles by using a test lung.
- The number of breaths delivered during a convenient interval must be counted by using a clock.
- Exhaled volume (tidal volume, sigh volume and minute volume) must be measured by using an external device such as a Wright respirometer or equivalent to independently measure exhaled volume.
- Ventilator sensitivity level must be checked in assist mode.
- Expose the oxygen monitor (or analyzer) used with the ventilator to room air and to wall oxygen (100%), and calibrate it.
- Ensure that a high-efficiency particulate-air (HEPA) filter is present on the main inspiratory line.

Numerous international standards are available for specifying the safety features of the anaesthesia machines [16–18]. Anaesthesia machines are covered by the ASTM (the American Society for Testing and Materials) standards. The most popular ones are:

- the ASA (the American Society of Anaesthesiologists),
- the CAS (Canadian Society of Anaesthesiologists)
- the ABZCA (Australian and New-Zealand College of Anaesthetists).

Most NGOs e.g. AAGBI (Association of Anaesthetists of Great Britain and Ireland) [19] and ASA recommend pre-anaesthesia test procedures that control the proper functioning of all the safety features incorporated in the machine.

Safety features of anaesthesia machine can be divided into the following units:

- **Gas supply unit**: In most of the new type devices, gas supply and monitoring systems can perform as high, intermediate and low-pressure systems. It must be verified from the central pipeline to the machine as well as all cylinders.
- **Flow meter unit**: The risk of oxygen loss because of the oxygen flow meter being positioned upstream from all other gases has been found in the reports. Therefore, the oxygen flow meter is always positioned downstream in a

sequence of flow meters in the modern devices. If there is a loss or leakage anywhere upstream of any other gas, still oxygen will be provided in a sufficient concentration to the unit.

- **Vaporizer unit**: A number of errors from the use of vaporizer units have been investigated i.e. filling wrong agents, wrong installation leading to loss of fresh gas flow, using multiple vaporizers simultaneously and filling gas channels with a liquid agent due to inappropriate transport arrangements. All these kind of potential risks have led to addition of safety mechanisms for vaporizers as mandated by the ISO and ASTM.
- **Fresh gas delivery unit for breathing systems and ventilators**: The entire system must be checked to satisfy leak free use and correct gas selection.
- **Scavenging**: It is an ignored part in the anaesthesia machine sometimes due to cost, ignorance, lack of health safety checks etc. Scavenging systems are handled both by ASTM and international standards. Scavenging systems also incorporate negative and positive pressure relief valves to make sure no dangerous pressures are transmitted into the breathing system in the event of malfunction of the system.

3.1 Comparison of IEC and All Other Manufacturers Testing Procedures and the Suggestion of a New One

Whenever measurements are realized, it is with the objective of generating data. The data is then processed, analysed and compared with requirements so that an appropriate decision can be taken. Reliability and accuracy of all those measurements and controls would be questionable if the devices used were not calibrated. Calibration guarantees that a measuring instrument displays an accurate and reliable value of the quantity being measured. Therefore, calibration is an indispensable action in any measurement process. A measurement must be traceable to the acceptable standard for it to be compared. A measurement result is meaningful only if it is presented with uncertainty value. Uncertainty is a measure of the quality of a measurement. It ensures the means to assess and minimise the risk and possible consequences of poor decisions.

International measurement community establishes documentary standards (procedures) that define how such quantities (flow, pressure etc.) are to be measured so as to provide the means for comparing the quality of goods or providing that safety and health requirements are satisfied. Therefore three elements are required in order to make a traceable measurement.

- Recognised and appropriate definition of how the quantity should be measured
- Calibrated measuring device
- Qualified and trained person who is able to evaluate the standard/procedure, and use the device systems.

The following routine controls should be carried out at the beginning of each operating session. In addition, specified controls should be carried out before each new patient during a session or when there is any change or addition to the breathing system, display or auxiliary equipment. It is the responsibility of the anaesthetist to make sure that all of these controls have been performed correctly.

- Perform machine controls given by manufacturer: Modern anaesthesia workstations may perform many of the following checks automatically during start-up.
- Power supply controls: Ensuring that the anaesthetic device and relevant auxiliary systems are connected to the main electrical supply and switched on.
- Gas supplies and suction controls: Checking that the correct function of the oxygen failure alarm covers the disconnection of the oxygen pipeline on some machines, whilst on machines with a gas supply master switch, the alarm may be operated by switching off the main switch.
- Medical gas supplies checks: Assign and notice all gases that are being.
- provided by a pipeline network, confirming with a—tug test—that each pipeline is properly and accurately inserted into the appropriate gas supply terminal.

  - Check that the anaesthetic device is connected to a supply of oxygen and that an sufficient reserve supply of oxygen exists from a spare cylinder.
  - Check that sufficient supplies of any other gases intended for use are provided and connected as convenient.
  - Check the mechanism and operation of flowmeters, where these are present, ensuring that each control valve smoothly operates.
  - Utilize the emergency oxygen bypass control and assure that flow occurs from the gas outlet without significant decrease in the pipeline supply pressure.
  - Check that the suction apparatus is functioning and all connections are secure.

- Breathing system and vaporizers controls: Check all breathing mechanisms that are to be used and implement a 'two-bag test' before use, as it is described below. Perform a pressure leak test on the breathing arrangement by occluding the patient-end and compressing the reservoir pocket. Manual leak testing of vaporizers was initially recommended routinely.
- Check that the vaporizer(s) for the required volatile agent(s) are properly fitted to the anaesthetic device. A manual leak test of vaporizer must be performed. Some anaesthetic workstations will automatically test vaporizer integrity. Check the CO_2 absorber and correct gas outlet.
- Ventilator control: Check that the ventilator is configured accurately for intended use. Ensure that the ventilator tubing is securely connected. Set the controls for use and guarantee that appropriate pressure is generated during the inspiratory phase.
- Two-bag test control: A two-bag test should be applied after the vaporizers, breathing system, and ventilator have been checked individually.

- Scavenging control: Examine that the anaesthetic gas scavenging mechanism is switched on and functioning adequately. Provide that the tubing is connected to the appropriate exhaust port of the ventilator, breathing system or anaesthetic workstation.
- Monitoring equipment tests: Inspect that all monitoring devices, particularly those referred to in the AAGBI's Standards of Monitoring during Anaesthesia and Recovery guidelines, are functioning and that appropriate parameters and alarms have been adjusted before using the anaesthetic system.
- Airway equipment control: Check all bacterial filters, catheter mounts, connectors etc.

ECRI (Emergency Care Research Institute) recommends a test procedure as a complete operational verification procedure must be described in a facility's policies/ procedures and applied periodically. It should usually be based on the manufacturer's recommendations as follows [20]:

- Verification of the volumetric flow rate(s),
- Checking the efficiency of the system periodically,
- Obtaining the specific information and comparing with design data,
- Setting a baseline for periodic maintenance controls,
- Principles for future installation design where satisfactory air contaminant control is currently being obtained,
- Meets regulatory or governmental requirements for particular types of processes.

Average air velocity must be measured by means of calculating $Q = V \cdot A$, where V is mean air velocity and A is average cross sectional area.

Three air pressures must be measured at any point in the exhaust system by means of calculating $TP = SP + VP$, where TP is total pressure, SP is static pressure and VP is velocity pressure in mmH_2O.

Devices used for measurements are the piezometer, U-tube manometer filled with oil or convenient liquid, water gauge and pressure gauge display. It should be noted that an inclined manometer yields increased accuracy and allows reading of lower velocity values.

Measurement methods include the measuring of hood static pressure by means of a U-tube manometer at one or more holes. The manometer is connected to each hole in turn by means of a thick walled soft rubber tube. The difference in height of the water columns is measured. After hood static pressure (SP_h) is known, the volumetric flow rate is determined as $Q = 4005 \cdot A \cdot C_e \cdot \sqrt{SP_h}$, where Q is flow rate in m^3/s. A is the average cross-sectional area in m^2 sqft, C_e is coefficient of entry loss and SP_h is static pressure in the hood or the duct.

In the velocity pressure and velocity of flow measurements the Pitot tube is used. This device consists of two concentric tubes. One measures the total or impact pressure existing in the air stream, while the other measures the static pressure only. The annular space and center tube are connected across a manometer. The velocity of air stream for standard conditions is determined as; $VPe = VPm/df$ where VPe is equivalent velocity pressure, VPm is measured velocity pressure and df is density correction factor.

In air velocity measurements the rotating vane anemometer, swinging vane anemometer, thermal anemometer, smoke tubes, tracer gas method and Pitot tubes are used. Rotating vane anemometer is used to determine the air flow through large supply and exhaust openings. It is used for either pressure or suction measurements in the range of 10–15 m^3/s. Swinging Vane Anemometer is used for field measurements. It is highly portable and has a wide scale range giving instantaneous readings. Thermal Anemometer consists of a velocity sensor and temperature sensor. The amount of heat removed by an air stream passing a heated object is related to the velocity of the air stream. Tracer gas is metered continuously into one or more intake ports along with entering air stream in the tracer gas method. Air samples are collected at some point downstream and the concentration of tracer gas in the exit stream is determined. The rate of air flow equals the rate of feed divided by tracer gas concentration.

Several international guidelines are available for anaesthesia machine check. The following protocol was developed based on the existing literature and individual practices, which involves the checking for the pneumatic, electrical, electronic and other components of the machine in a systematic manner [21–23].

The following recommendations are aimed at providing basic guidelines to anaesthetic practice. They are intended to provide a framework for reasonable and acceptable patient care and should be so interpreted, allowing for some degree of flexibility in different circumstances. Each section of these guidelines is subject to revision as warranted by the evolution of technology and practice.

3.2 Safety Testing of Ventilators According to IEC 60601 Standard

An anaesthetic machine is a continuous flow machine that is designed to provide accurate and continuous supply of medical gases (such as oxygen and nitric oxide), in mixtures with a precise concentration of anaesthetic vapour (such as isoflurane and sevoflurane (Sevoran)), and deliver it to the patient at controlled pressure and flow.

Respirators and anaesthesia machines must be designed and constructed so that it protects against electric shock, excessive temperature, dust and water at under normal operating conditions. All parts of respirators and anaesthesia machines that are subject to normal working conditions. Corrosion must be effectively protected. This protection must not be susceptible to damage to handling. Respirators and anaesthetic machines must have a name plate on which they are visible and printed in such a way that cannot be deleted or removed during normal use. The name plate shall contain the following elements:

- The manufacturer's name or label
- Serial number and year of production
- Type markers
- Metal type designation.

Respirators and anaesthesia machines must undergo the procedure of type approval testing and have type approval certificates.

Respirators and anaesthesia machines must meet the following metrological and technical requirements. The verification periods are defined by the regulations. The manufacturer must ensure that these instruments can be used under reference conditions.Reference conditions for the respirator are:

- Input voltage: (100–240) V AC, 50/60 Hz
- (12–24) V DC internal battery (when battery operated)

  - Charging time: < 6 h
  - The battery life of the respirator and compressor is at least 30 min
  - The operation of the battery for the respirator is maximally 7 h
  - The battery life of the compressor and respirator is maximally 2 h.

- Concentration of air and oxygen volumes: (18–100)%,
- Current air pressure (if the value is entered manually).

Reference conditions for anaesthesia machines are:

- Input voltage: (100–240) V AC, 50/60 Hz
- Availability of internal rechargeable battery

  - Working time is minimum 30 min.

- Ambient operating conditions of the system:

  - Temperature: (10–40) °C,
  - Relative humidity: (15–95)% rh.

Concentrations of anaesthetic gases are:

- CO_2: (0–20)%
- NO_2: (0–100)%
- HAL, ISO, ENF: (0–12)%
- SEV: (0–15)%
- DES: (0–22)%

The measuring ranges for the respirator and an anaesthetic machine are as follows:

- Low flow: (− 60 to 40) L/min
- High flow: (− 300 to 200) L/min
- Output pressure of the respirator: (− 60 to 140) cmH_2O
- Volume: (− 1.00 to 4.00) L.

All measurements must comply with the requirements and guidelines of the standards EN IEC 60601-1 "General Requirements for Electrical Medical Equipment" [24].

Measurements errors cannot be more than the stated values below:

- Flow rate: ± 10% of reading
- Output pressure of the respirator: ± 5% reading
- Volume: ± 10% of the reading
- Deviation of concentration in anaesthetic gases: ± 1% of the reading value.

Certificate of verification must be given and include:

- General, technical and other documentation related to compliance with other standards, which allows conformity of the type of measurement.
- Instructions for use which must include more information and description of the benchmarks with all its parts.
- Information and operation of the software, if the device is equipped with a microprocessor.

4 Summary

Several respirator performance criteria are set to satisfy the physiological requirements of the worker. Filtration principles and the nature of workplace aerosols must also be understood to determine appropriate test conditions for particulate respirator filters. A current filter test criterion assures that significant aerosol penetration will not occur in the workplace.

As can be seen, a respirator program is not a simple task, especially if the camp does not have the expertise to evaluate all of the influencing factors. However, it is an important task if the exposure to employees exists. It is perhaps best addressed for most by consulting the regional OSHA office or by using local, qualified vendors who can assist in the selection and fitting of respirators as well as in establishing appropriate respirator-related procedures and protocols.

The review highlights the fact that problems can occur despite the incorporation of several safety aspects to an anaesthesia machine. Human factors have contributed to greater complications than machine faults. Therefore, better understanding of the basics of the anaesthesia machine and checking each component of the machine for proper functioning prior to use is essential to minimise these hazards. Despite advanced technology, a remote but life-threatening possibility of intraoperative machine malfunction exists. A self-inflating bag appropriate for the patient's age and alternate O_2 source should be present as rescue measures in the event of machine malfunction.

Standard

No	Standard
1	ISO 8185:2007: Respiratory tract humidifiers for medical use—Particular requirements for respiratory humidification systems
2	ISO 9360-1:2000: Anaesthetic and respiratory equipment—Heat and moisture exchangers (HMEs) for humidifying respired gases in humans—Part 1: HMEs for use with minimum tidal volumes of 250 ml
3	ISO 9360-2:2001: Anaesthetic and respiratory equipment—Heat and moisture exchangers (HMEs) for humidifying respired gases in humans—Part 2: HMEs for use with tracheostomized patients having minimum tidal volumes of 250 ml
4	ISO 10651-3:1997: Lung ventilators for medical use—Part 3: Particular requirements for emergency and transport ventilators
5	ISO 10651-4:2002: Lung ventilators—Part 4: Particular requirements for operator-powered resuscitators
6	ISO 10651-5:2006: Lung ventilators for medical use—Particular requirements for basic safety and essential performance—Part 5: Gas-powered emergency resuscitators
7	ISO 10651-6:2004: Lung ventilators for medical use—Particular requirements for basic safety and essential performance—Part 6: Home-care ventilatory support devices
8	ISO/TR 13154:2009: Medical electrical equipment—Deployment, implementation and operational guidelines for identifying febrile humans using a screening thermograph
9	ISO/PRF TR 13154: Medical electrical equipment—Deployment, implementation and operational guidelines for identifying febrile humans using a screening thermograph
10	ISO 17510:2015: Medical devices—Sleep apnoea breathing therapy—Masks and application accessories
11	ISO/FDIS 18562-1: Biocompatibility evaluation of breathing gas pathways in healthcare applications—Part 1: Evaluation and testing within a risk management process
12	ISO/FDIS 18562-2: Biocompatibility evaluation of breathing gas pathways in healthcare applications—Part 2: Tests for emissions of particulate matter
13	ISO/FDIS 18562-3: Biocompatibility evaluation of breathing gas pathways in healthcare applications—Part 3: Tests for emissions of volatile organic compounds (VOCs)
14	ISO/FDIS 18562-4: Biocompatibility evaluation of breathing gas pathways in healthcare applications—Part 4: Tests for leachables in condensate
15	ISO 18778:2005: Respiratory equipment—Infant monitors—Particular requirements
16	ISO/DIS 20789: Anaesthetic and respiratory equipment—Passive humidifiers
17	ISO 23328-1:2003: Breathing system filters for anaesthetic and respiratory use—Part 1: Salt test method to assess filtration performance
18	ISO 23328-2:2002: Breathing system filters for anaesthetic and respiratory use—Part 2: Non-filtration aspects
19	ISO 23747:2015: Anaesthetic and respiratory equipment—Peak expiratory flow meters for the assessment of pulmonary function in spontaneously breathing humans
20	ISO 26782:2009: Anaesthetic and respiratory equipment—Spirometers intended for the measurement of time forced expired volumes in humans

(continued)

(continued)

No	Standard
21	IEC/NP 60601-2-83: Medical electrical equipment—Part 2–83: Particular requirements for the basic safety and essential performance of home light therapy equipment
22	IEC 60601-1-8:2006: Medical electrical equipment—Part 1–8: General requirements for basic safety and essential performance—Collateral standard: General requirements, tests and guidance for alarm systems in medical electrical equipment and medical electrical systems
23	IEC 60601-1-10:2007: Medical electrical equipment—Part 1–10: General requirements for basic safety and essential performance—Collateral standard: Requirements for the development of physiologic closed-loop controllers
24	IEC 60601-1-11:2015: Medical electrical equipment—Part 1–11: General requirements for basic safety and essential performance—Collateral standard: Requirements for medical electrical equipment and medical electrical systems used in the home healthcare environment
25	IEC 60601-1-12:2015: Medical Electrical Equipment—Part 1–12: General requirements for basic safety and essential performance—Collateral Standard: Requirements for medical electrical equipment and medical electrical systems used in the emergency medical services environment
26	IEC/CD 60601-2-26: Medical electrical equipment—Part 2–26: Particular requirements for the basic safety and essential performance of electroencephalographs
27	IEC/DIS 60601-2-49: Medical electrical equipment—Part 2–49: Particular requirements for the basic safety and essential performance of multifunction patient monitoring equipment
28	ISO 80601-2-12:2011: Medical electrical equipment—Part 2–12: Particular requirements for basic safety and essential performance of critical care ventilators
29	ISO/CD 80601-2-12: Medical electrical equipment—Part 2–12: Particular requirements for basic safety and essential performance of critical care ventilators
30	IEC/DIS 80601-2-30: Medical electrical equipment—Part 2–30: Particular requirements for basic safety and essential performance of automated non-invasive sphygmomanometers
31	IEC 80601-2-30:2009: Medical electrical equipment—Part 2–30: Particular requirements for basic safety and essential performance of automated non-invasive sphygmomanometers
32	ISO/FDIS 80601-2-56: Medical electrical equipment—Part 2–56: Particular requirements for basic safety and essential performance of clinical thermometers for body temperature measurement
33	ISO 80601-2-56:2009: Medical electrical equipment—Part 2–56: Particular requirements for basic safety and essential performance of clinical thermometers for body temperature measurement
34	IEC 80601-2-59:2008: Medical electrical equipment—Part 2–59: Particular requirements for basic safety and essential performance of screening thermographs for human febrile temperature screening
35	IEC/DIS 80601-2-59: Medical electrical equipment—Part 2–59: Particular requirements for the basic safety and essential performance of screening thermographs for human febrile temperature screening
36	ISO/DIS 80601-2-61: Medical electrical equipment—Part 2–61: Particular requirements for basic safety and essential performance of pulse oximeter equipment

(continued)

(continued)

No	Standard
37	ISO 80601-2-61:2011: Medical electrical equipment—Part 2–61: Particular requirements for basic safety and essential performance of pulse oximeter equipment
38	ISO 80601-2-67:2014: Medical electrical equipment—Part 2–67: Particular requirements for basic safety and essential performance of oxygen-conserving equipment
39	ISO 80601-2-69:2014: Medical electrical equipment—Part 2–69: Particular requirements for basic safety and essential performance of oxygen concentrator equipment
40	ISO 80601-2-70:2015: Medical electrical equipment—Part 2–70: Particular requirements for basic safety and essential performance of sleep apnoea breathing therapy equipment
41	IEC 80601-2-71:2015:Medical electrical equipment—Part 2–71: Particular requirements for the basic safety and essential performance of functional Near-Infrared Spectroscopy (NIRS) equipment
42	ISO 80601-2-72:2015: Medical electrical equipment—Part 2–72: Particular requirements for basic safety and essential performance of home healthcare environment ventilators for ventilator-dependent patients
43	ISO/DIS 80601-2-74: Medical electrical equipment—Part 2–74: Particular requirements for basic safety and essential performance of respiratory humidifying equipment
44	ISO/CD 80601-2-79: Medical electrical equipment—Part 2–79: Particular requirements for basic safety and essential performance of home healthcare environment ventilatory support equipment for respiratory impairment
45	ISO/CD 80601-2-80: Medical electrical equipment—Part 2–80: Particular requirements for basic safety and essential performance of home healthcare environment ventilatory support equipment for respiratory insufficiency
46	ISO 81060-1:2007: Non-invasive sphygmomanometers—Part 1: Requirements and test methods for non-automated measurement type
47	ISO 81060-2:2013: Non-invasive sphygmomanometers—Part 2: Clinical investigation of automated measurement type
48	ISO/NP 81060-3: Non-invasive sphygmomanometers—Part 3: Clinical investigation of continuous automated measurement type
49	ISO 5362:2006: Anaesthetic reservoir bags
50	ISO 8185:2007: Respiratory tract humidifiers for medical use—Particular requirements for respiratory humidification systems
51	ISO 9360-1:2000: Anaesthetic and respiratory equipment—Heat and moisture exchangers (HMEs) for humidifying respired gases in humans—Part 1: HMEs for use with minimum tidal volumes of 250 ml
52	ISO 23328-2:2002: Breathing system filters for anaesthetic and respiratory use—Part 2: Non-filtration aspects
53	ISO 4135:2001: Anaesthetic and respiratory equipment—Vocabulary
54	ISO 5367:2014: Anaesthetic and respiratory equipment—Breathing sets and connectors

References

1. Geddes LA (2007) The history of artificial respiration. IEEE Eng Med Biol Mag: Q Mag Eng Med Biol Soc 26(6): 38–41. doi:10.1109/EMB.2007.907081.PMID 18189086
2. Russell WR, Schuster E, Smith AC, Spalding JM (1956) Radcliffe respiration pumps. The Lancet. 270(6922): 539–541. doi:10.1016/s0140-6736(56)90597-9. PMID 13320798
3. Bellis M (2009) Forrest bird invented a fluid control device, respirator and pediatric ventilator. www.about.com. Retrieved 4 June 2009
4. OSHA Office of Training and Education, Training and Reference Materials Library (2004) Respiratory protection frequently asked questions, Rev. Nov 2004. https://www.osha.gov/dte/library/respirators/faq.html
5. Brockwell RC, Andrews GG (2002) Understanding your anaesthesia machine. ASA refresher courses, vol 4. Lippincott Williams & Wilkin, Philadelphia, pp 41–59
6. Thompson PW, Wilkinson DJ (1985) Development of anaesthetic machines. Br J Anaesth 57:640–648
7. Dorsch JA, Dorsch SE (eds) (2007) Understanding anesthesia equipment. 5th edn. Hazards of anesthesia machines and breathing systems, Lippincott Williams and Wilkins, Philadelphia, pp 373–403
8. Brockwell RC, Andrews JJ (2002) Complications of inhaled anesthesia delivery systems. Anesthesiol Clin North America. 20:539–554
9. Sugiuchi N, Miyazato K, Hara K, Horiguchi T, Shinozaki K, Aoki T (2000) Failure of operating room oxygen delivery due to a structural defect in the ceiling column. Masui. 49:1165–1168
10. Aitkenhead AR (2005) Injuries associated with anaesthesia. A global perspective. Br J Anaesth 95(1):95–109. Epub 2005 May 20
11. Posner KL (2001) Closed claims project shows safety evolution. Anesthesia Patient Safety Foundation Newsletter
12. FDA, MAUDE—manufacturer and user facility device experience database. http://www.accessdata.fda.gov/scripts/cdrh/cfdocs/cfMAUDE/Search.cfm?smc=1
13. Subrahmanyam M, Mohan S (2013) Safety features in anaesthesia machine, Indian J Anaesth 57(5): 472–480. doi:10.4103/0019-5049.120143
14. Skinner M (1998). Ventilator function under hyperbaric conditions. South Pac Underw Med Soc J 28(2). Retrieved 4 June 2009
15. Weaver LK, Greenway L, Elliot CG (1988). Performance of the Seachrist 500A hyperbaric ventilator in a monoplace hyperbaric chamber. J Hyperb Med 3(4): 215–225. Retrieved 4 June 2009
16. Anderson HE, Bythell V, Gemmell L, Jones H, McIvor D, Pattinson A, Sim P, Walker I (2012) Checking anaesthetic equipment. Anaesthesia 67: 660–668. doi:10.1111/j.1365-2044.2012.07163.x]
17. Cassidy CJ, Smith A, Arnot-Smith J (2011) Critical incident reports concerning anaesthetic equipment: analysis of the UK national reporting and learning system (nrls) data from 2006–2008. Anaesthesia 66:879–888
18. Merchant R, Chartrand D, Dain S, Dobson G, Kurrek M, Lagace A, Stacey S, Thiessen B (2013) Guidelines to the practice of anaesthesia Revised Edition 2013
19. Association of Anaesthetists of Great Britain and Ireland. Standards of Monitoring during Anaesthesia and Recovery 4, 2007. http://www.aagbi.org/publications/guidelines/docs/standards ofmonitoring07.pdf. Accessed 29 Jan 2012)
20. ECRI Institute (1998) Medical device safety reports, minimum requirements for ventilator testing, guidance, health devices 27(9–10):363–364 http://www.mdsr.ecri.org/summary/detail.aspx?doc_id=8305. Guidance [Health Devices Sep-Oct 1998;27(9-10):363-4]
21. Goneppanavar U, Prabhu M (2013) Anaesthesia machine: checklist, hazards, scavenging. Indian J Anaesth 57(5): 533–540. doi:10.4103/0019-5049.120151. https://www.ncbi.nlm.nih.gov/pmc/articles/PMC3821271/

22. Badnjevic A, Gurbeta L, Jimenez ER, Iadanza E (2017) Testing of mechanical ventilators and infant incubators in healthcare institutions. Technol Health Care J
23. Gurbeta L, Badnjevic A, Sejdinovic D, Alic B, Abd El-Ilah L, Zunic E (2016) Software solution for tracking inspection processes of medical devices from legal metrology system. XIV Mediterranean conference on medical and biological engineering and computing (MEDICON), Paphos, Cyprus, 31 Mar–02 April
24. IEC EN 60601-1 General requirements for electrical medical equipment

Inspection and Testing of Dialysis Machines

Dušanka Bošković

Abstract Dialysis machines act as artificial kidney performing extracorporeal blood purification to remove excess water, detoxify the blood and balance the blood composition. The chapter presents development of dialysis from experiments to regular life-saving clinical practice, leading to modern dialysis machine organization and functionality. Function deterioration of a hemodialysis device, as a potential harm for a patient safety, is discussed. Further on, overview of standards related to hemodialysis machines safety and risk management is presented. The chapter concludes with a description of a procedure for safety performance inspection focused on key dialysis parameters: temperature and conductivity of dialysate, and blood pressure.

The dialysis machine is a therapeutic device aimed to provide hemodialysis treatment for patients with renal failure. Dialysis machines have central role as renal replacement therapy, with 2 million people in the world receive treatment either with dialysis or a kidney transplant. It is estimated that this number represent only 10% of people who need treatment to live [1].

In the course of the hemodialysis the blood is taken from the shunt between the veins and arteries of the patient forearm. The blood is circulated into the dialyzer for excess water and uremic toxins removal; and returned to the patient. Hemodialysis parameters as flow, temperature, blood pressure, blood leaks, etc. are monitored and controlled during the treatment.

The developments in design and construction of dialysis machines lead to contemporary devices (Fig. 1) equipped with controls, monitors, and alarms that provide for safe proportioning of dialysate.

D. Bošković (✉)
Faculty of Electrical Engineering, University of Sarajevo, Sarajevo
Bosnia and Herzegovina
e-mail: dboskovic@etf.unsa.ba

© Springer Nature Singapore Pte Ltd. 2018
A. Badnjević et al. (eds.), *Inspection of Medical Devices*, Series in Biomedical Engineering, https://doi.org/10.1007/978-981-10-6650-4_10

(a)   **(b)**

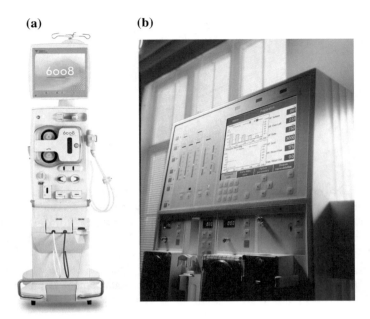

Fig. 1 Hemodialysis 6008 machine (**a**), and online clearance monitor for the 4008H/S (**b**), courtesy Fresenius Medical Care

1 History of Hemodialysis

The process of dialysis is based on principles of osmosis and diffusion. Thomas Graham (1805–1869), a chemist at the Glasgow University, conducted experiments on separating substances using a semi-permeable membrane made of bladders. Graham described the process in 1854 in his paper "Bakerian lecture on osmotic forces" [2]. He noticed the importance of controlling the rates of transfer through the membrane for successful removal of toxins from the blood, and repeated his experiments with different solutes to measure and compare the rates. Graham was the first to use parchment membrane, what he described in paper published in 1861 and he was the first to apply the term dialysis "*to the method of separation by diffusion through a septum of a gelatinous matter*" [3].

In 1855 Adolf Fick (1829–1901) provided mathematical description for selective transport processes through a semipermeable membrane caused by concentration gradients [4]. The description would became known as Fick's laws of diffusion. After Fick, who was the first to use collodion membranes, Schumacher in 1860 obtained excellent results with the skins from collodion, and afterwards with collodion tubes [5]. It was not until 1921 that Arnold Eggerth described standardized process for preparation of collodion membranes with controlled difference in water permeability [6].

The first dialysis device was developed in 1914 at Johns Hopkins University School of Medicine by John Abel (1857–1938) and his colleagues, and they named the device "artificial kidney" [5]. Their device used collodion tubes similar to present day hollow fiber dialyzers and was not used on humans, only for experiments on animals. The first hemodialysis device used to treat human patient was performed by George Haas in 1924 [7]. He developed a glass cylinder dialyzer with collodion tubes, and the first treatment lasted for 15 min. Hass performed several hemodialysis procedures in the following years and reported first clinical results. In his work for producing collodion membranes he followed the standards and procedures described by Eggerth [7].

The pioneer of artificial organs, Willem Kolff (1911–2009) from Netherlands, built in 1943 the rotating drum hemodialysis device using cellophane membranes. Kolff was treating patients with the acute renal failure in the following years, and after the series of unsuccessful treatments, in 1945 his hemodialysis treatment saved a patient's life. Kolff continued his work in the USA and he was influential in developing hemodialysis from experimental into standard clinical therapy.

In 1947 Swedish scientist Nils Alwill (1904–1986) described his apparatus for dialysis with improved function for removal of excess water, based on combination of dialysis and ultrafiltration. Alwill was using cellophane membrane with tight fitting container enabling better control of extracorporeal blood volume [7].

Hemodialysis was not seen as a solution for end-stage chronic renal failure, since dialysis treatment damaged patients' veins and arteries. Typical patient with chronic renal failure would need dialysis treatment three times a week, with sessions lasting several hours, so the next important challenge for hemodialysis treatment was development of a suitable vascular access. Scribner and colleagues created in 1960 a shunt between the radial artery and the cephalic vein using an external silicon device [8]. This solution enabled hemodialysis treatments for patients with end-stage renal failure and marked inception of dialysis as a standard clinical treatment. In 1966 Breschia and Cimino proposed a surgical arteriovenous (AV) fistula as a solution to eliminate the external shunt prone to bleeding and infection. AV fistula demonstrated as safe and long-lasting vascular access [8].

The hemodialysis developed significantly since 1960 with modern dialysis machines with improved dialyzer design, volumetric control, embedded monitoring and alarming systems, high flux membranes. In order to ensure no chemical or bacterial contamination in dialysate, important requirement for dialysis membrane is adsorption capability for bacterial and organic contaminants and inflammatory mediators. In addition, membrane interaction with blood requires blood compatibility and biocompatibility [9]. Significant developments over the recent years are online dialysis monitoring devices and continuous therapy. The modern hemodialysis devices continually measure and record values allowing features as: sodium profiling, ultrafiltration variation based on blood pressure measurement, urea kinetics, etc. [10].

Although dialysis devices are still limited in their functions compared with the complex physiological tasks of the natural kidneys, improvement trends are focused on: maximum removal of uremic retention substances combined with reduced patient exposure to influence of inflammatory stimuli [9].

2 Dialysis Machine Organization and Functionality

The major components of the hemodialysis machine are: (a) extracorporeal blood circuit, (b) dialysate circuit, and (c) dialyzer. The hemodialysis machine provide monitoring and control functions in order to provide safe, efficient and accurate treatment. The basic schematic diagram of the dialysis machine with the relevant parts is presented in Fig. 2.

Basic functions of dialysis machine are: dialysate preparation, dialysate heating to physiological range, monitoring of conductivity and pH, controlling fluid removal, pumping blood and anticoagulant at determined rates, and monitoring pressure in the extracorporeal blood circuit.

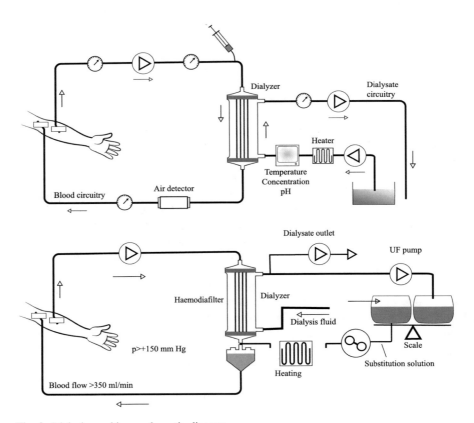

Fig. 2 Dialysis machine—schematic diagram

The *extracorporeal blood circuit* main function is to provide the continuous blood circulation from the patient through a dialyzer, where waste is removed through a semi-permeable membrane, and returned to the patient. The hemodialysis machine administer anticoagulant (heparin) throughout the treatment. The parameters monitored in the blood circuitry are venous and arterial blood pressures, and additional control is air presence detection.

The *dialysate circuitry* is responsible to deliver dialysate fluid and maintain its appropriate temperature, pressure and concentration. Main components of dialysate circuitry are: heating, deaeration, proportioning, ultrafiltration, and monitoring. Monitoring is used for continual verification of the composition of the dialysate and to detect any abnormal occurrence as blood leak in dialysate circuit.

The blood purification takes place in the *dialyzer*, as presented in Fig. 3. Dialyzer contains about 20,000 hollow membranes imitating the filtering function of the natural kidney unit: the glomerulus. The capillary membranes have an inner diameter of 200 μm. The wall of the capillaries consist of membranes with a thickness of about 40 μm through which material exchange takes place [9].

The space between the capillary membranes are perfused with a dialysis fluid or dialysate in opposite flow to the direction of blood flow. This creates a concentration gradient and various uremic toxins are transferred out of the blood into the dialysate as filtrate, by means of the concentration gradients across the semipermeable capillary membranes. The mass transfer of the solutes is based on principles of diffusion (conduction transfer) and ultrafiltration (convection transfer) [11]. A measure for the efficiency of dialysis is clearance, defined as amount of substance removed from blood over a unit of time, divided by the respective concentration in the blood of a patient [9]. Clearance is calculated as a sum of diffusive and convective portion.

Monitoring devices are responsible to detect and identify hazardous events potentially harmful to patients such as: blood leaks, incorrectly dialysate pressures or temperature, and air in the blood. After detecting adverse event device should trigger an alarm and stop circulation in the blood and/or the dialysate circuitry.

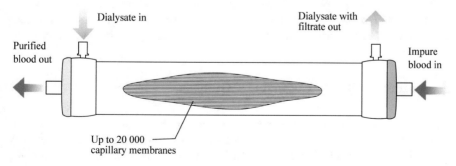

Fig. 3 Dialyzer

Control of the blood pump in the blood circuitry is related to the following:

- Monitor for continual measuring of the blood pressure in the extracorporeal blood circuit in the venous segment.
- Air detection in the venous segment before blood is returned into vein.
- Blood leak monitoring by method of detecting hemoglobin in the dialysate circuitry.

Control of the dialysate circuitry is related to monitoring the following parameters:

- Temperature sensor provides a short feedback to heater to maintain body temperature (35–39) °C; low temperatures can cause shivering and high temperatures can cause protein denaturing or hemolysis.
- Conductivity monitor checks the electrolyte concentrations and prevents any adverse events caused by incorrect dialysate proportioning.
- Measuring pH value, as useful additional parameter for the dialysate proportioning monitoring.

If the concentration of electrolytes changes, the voltage will change, thus increase in electrolytes shall increase the conductivity of the dialysate. Conductivity reflects electrolyte concentration of dialysate and provides for real-time estimate of concentration [12, 13]. This feature of modern dialysis machines support comparison with other important physiological parameters and can help in deciding on appropriate composition of the dialysate what has been one of the central topics in the delivery of dialysis treatment [14]. Alarm limits for conductivity can be adjusted based on set-up concentrate composition and for standard temperature. Conductivity measurements are temperature compensated. Although monitoring display range can be as 10–17 mS/cm, devices have predefined safety alarm limits, as e.g. 12–16.5 mS/cm [15]. Conductivity is also used for modelling sodium mass transfer; calculating differences in conductivity values measured pre and post dialyzer, and conductivity can be used as a surrogate for sodium concentration with one mS/cm conductivity equivalent to 10 meq/L sodium [12].

During the design of a medical device, potential risks and hazards should be identified, and monitoring, control and alerting functions foreseen in order to reduce probability of occurrence of any harm to patient health to a minimum level.

3   Potential Safety Hazards in Dialysis

Safety of a patient is the most important requirement for medical device design and operation. Erroneous hemodialysis device or improper clinical application may cause different serious adverse effects as hemorrhage, low blood pressure, uremia, and etc. [16–19].

A multidisciplinary task force composed by engineers representing manufacturers of dialysis machines, nephrology clinicians, and dialysis experts published their thorough analysis of hazards and harms related to hemodialysis devices [20].

Based on reviewing adverse events databases, literature and clinical experience, the task force identified and described more than 50 different harms related to a hemodialysis device. The harms are classified in five levels of severity and linked to underlying hazards. The hazards are classified as: biocompatibility, biological, chemical, electromagnetic, mechanical, function deterioration, use error and labelling. Further on hazards are related to physical quantities if adverse event is caused by fault measurement.

Medical device inspections are aimed to prevent device function deterioration as a potential cause of adverse events and harm to a patient. Subset of harms identified in [20] is presented in Table 1, with a following selection criteria: hazards of a type function deterioration, including only hazards related to measurable quantity. With exception of Under dialysis, selected harms are classified with a highest level of severity (5 and 4).

Analyzing quantities in the Table 1, and knowing that conductivity reflects electrolyte concentration, it is possible to conclude that the key parameters for a patient safety are: dialysate temperature, dialysate concentration and extracorporeal blood pressure.

Table 1 Hemodialysis device hazards related to function deterioration

Harm	Hazardous situation	Related quantity	Severity (min = 1, max = 5)
Acid-base imbalance	High/low bicarbonate in dialysate/substitution fluid	Bicarbonate in dialysate (mmol/L)	4
Hemolysis	Reduced dialysate tonicity	Dialysate conductivity (mS/cm)	5
Hemolysis	Blood exposed to high temperature	Dialysate temperature (°C)	5
Hemolysis	Mechanical stress to red cells as a result of extracorporeal circulation	Extracorporeal BP (negative and positive; mmHg)	5
Hyperthermia	Extracorporeal blood exposed to high temperature	Dialysate temperature (°C)	5
Hypothermia	Extracorporeal blood exposed to low temperature	Dialysate temperature (°C)	5
Plasma electrolyte imbalance	High/low sodium in dialysate/substitution fluid	Sodium in dialysate (mmol/L)	5
Plasma electrolyte imbalance	High/low potassium in dialysate/substitution fluid	Potassium in dialysate (mmol/L)	5
Plasma electrolyte imbalance	High/low calcium in dialysate/substitution fluid	Calcium in dialysate (mmol/L)	5
Underdialysis	Reduced dialysis effectiveness (i.e., inadequate urea removal)	Kt/V	2

4 Standards and Regulations for Dialysis Machines

With an objective to protect hemodialysis patients from adverse effects, responsible international and national organizations provide standards, regulations and guidelines to the medical equipment manufacturers, and to the medical professionals for the use, care, and/or processing of a medical device or system.

4.1 Standards

The International Electrotechnical Commission (IEC) has provided the standard IEC 60601-1:2005 (Medical electrical equipment—Part 1) specifying general requirements for basic safety and essential performance of medical devices [21].

IEC 60601 is a series of technical standards for the safety and essential performance of medical electrical (ME) equipment. The consolidated version is the third edition IEC 60601-1 from 2005 with its amendment 1 from 2012. These general requirements are supplemented by the special requirements of collateral and particular standards.

IEC 60601-2-16:2012 is particular standard for basic safety and essential performance of hemodialysis, hemodiafiltration and hemofiltration equipment. This particular standard takes into consideration the specific safety requirements concerning electrical safety and patient safety, and does not take into consideration the dialysis fluid control system of hemodialysis equipment using regeneration of dialysis fluid and central delivery systems [22].

The general standard, IEC 60601 Part 1 is adopted in many countries by their regulatory bodies and recognized as requirement for the commercialization of medical equipment. Some countries also adopted standards or guidelines for further safety testing, inspection and calibration of medical equipment specific for different phases of the medical equipment life cycle. Some examples are: acceptance test, inspections performed in regular intervals, and/or directly following service or repair.

Regulations stipulating technical and measurement requirements for medical equipment include following chapters: terms and definitions, general technical requirements, measurement requirements, identification, marking and documents.

The IEC 60601 defines that ME equipment shall be marked with the information on supply voltage(s) or voltage range(s) to which it may be connected. ME equipment shall be accompanied by documents containing the instructions for use and a technical description, and also all data that is essential for safe operation, transport and storage, and measures or conditions necessary for installing the ME equipment, and preparing it for use.

It is a common practice that the IEC 60601 standard has been adapted into local national standard for use in specific country. In such cases national standards can introduce additional requirements specific for that environment, such as voltage to which the ME equipment may be connected.

Standard documents provide applicable terms and definitions. Dialysis machine is considered as ME equipment or ME system used to perform hemodialysis, hemofiltration and/or hemodiafiltration.

The IEC 60601-2-16 provide following definitions:

Hemodialysis (HD) is defined as "process whereby concentrations of water-soluble substances in a patient's blood and an excess of fluid of a patient with renal insufficiency are corrected by bidirectional diffusive transport and ultrafiltration across a semi-permeable membrane separating the blood from the dialyzing fluid."

Hemofiltration (HF) is defined as "process whereby concentrations of water-soluble substances in a patient's blood and an excess of fluid of a patient with renal insufficiency are corrected by unidirectional convective transport via ultrafiltration across a semi-permeable membrane separating the blood from the dialyzing fluid. Ultrafiltrate is simultaneously replaced by an approximately isoosmolar substitution fluid at a rate such that the difference between the ultrafiltration rate and the rate of substitution fluid addition will lead to removal of the excess fluid over the course of the treatment."

Hemodiafiltration (HDF) is defined as "process whereby concentrations of water-soluble substances in a patient's blood and an excess of fluid of a patient with renal insufficiency are corrected by simultaneous combination of HD and HF."

The scope of the IEC 60601-2-16 states that the particular requirements in that standard do not apply to: extracorporeal circuits, dialyzers, dialysis fluid concentrates, water treatment equipment, and equipment used to perform peritoneal dialysis.

International Standard Organization (ISO) provided several standards related to specific aspect of hemodialysis:

- ISO 8637:2010 Cardiovascular implants and extracorporeal systems— Hemodialysers, hemodiafilters and hemoconcentrators
- ISO 8638:2010 Cardiovascular implants and extracorporeal systems— Extracorporeal blood circuit for hemodialysers, hemodiafilters and hemofilters
- ISO 11663:2014 Quality of dialysis fluid for hemodialysis and related therapies
- ISO 13958:2014 Concentrates for hemodialysis and related therapies
- ISO 13959:2014 Water for hemodialysis and related therapies
- ISO 23500:2014 Guidance for the preparation and quality management of fluids for hemodialysis and related therapies
- ISO 26722:2014 Water treatment equipment for hemodialysis applications and related therapies

The following sections cite Abstracts of these standards:

- "ISO 8637:2010 specifies requirements for hemodialysers, hemodiafilters, hemofilters and hemoconcentrators for use in humans" [23].

- "ISO 8638:2010 specifies requirements for hemodialysers, hemodiafilters, hemofilters and hemoconcentrators and (integral and non-integral) transducer protectors which are intended for use in hemodialysis, hemodiafiltration and hemofiltration" [24].
- "ISO 11663:2014 specifies minimum quality requirements for dialysis fluids used in hemodialysis and hemodiafiltration, including substitution fluid for hemodiafiltration and hemofiltration" [25].
- "ISO 13958:2014 specifies minimum requirements for concentrates used for hemodialysis and related therapies. For the purpose of ISO 13958:2014, "concentrates" are a mixture of chemicals and water, or chemicals in the form of dry powder or other highly concentrated media that are delivered to the end user to make dialysis fluid used to perform hemodialysis and related therapies. ISO 13958:2014 is addressed to the manufacturer of such concentrates. ISO 13958:2014 includes concentrates in both liquid and powder forms. Also included are additives, also called spikes, which are chemicals that may be added to the concentrate to increase the concentration of one or more of the existing ions in the concentrate and thus in the final dialysis fluid. ISO 13958:2014 also gives requirements for equipment used to mix acid and bicarbonate powders into concentrate at the user's facility" [26].
- "ISO 13959:2014 specifies minimum requirements for water to be used in haemodialysis and related therapies. ISO 13959:2014 includes water to be used in the preparation of concentrates, dialysis fluids for hemodialysis, hemodiafiltration and hemofiltration, and for the reprocessing of hemodialysers" [27].
- "ISO 23500:2014 provides dialysis practitioners with guidance on the preparation of dialysis fluid for hemodialysis and related therapies and substitution fluid for use in online therapies, such as hemodiafiltration and hemofiltration. ISO 23500:2014 functions as a recommended practice. ISO 23500:2014 addresses the user's responsibility for the dialysis fluid once the equipment used in its preparation has been delivered and installed. For the purposes of ISO 23500:2014, the dialysis fluid includes dialysis water used for the preparation of dialysis fluid and substitution fluid, dialysis water used for the preparation of concentrates at the user's facility, as well as concentrates and the final dialysis fluid and substitution fluid" [28].
- "ISO 26722:2014 is addressed to the manufacturer and/or supplier of water treatment systems and/or devices used for the express purpose of providing water for haemodialysis or related therapies. ISO 26722:2014 covers devices used to treat water intended for use in the delivery of haemodialysis and related therapies, including water used for: (1) the preparation of concentrates from powder or other highly concentrated media at a dialysis facility; (2) the preparation of dialysis fluid, including dialysis fluid that can be used for the preparation of substitution fluid; (3) the reprocessing of dialysers for multiple uses. Included within the scope of ISO 26722:2014 are all devices, piping and fittings between the point at which potable water is delivered to the water treatment system, and the point of use of the dialysis water. Examples of devices included within the scope of ISO 26722:2014 are water purification devices,

online water quality monitors (such as conductivity monitors), and piping systems for the distribution of dialysis water" [29].

Other relevant ISO standards are:

- "ISO 13485:2016 specifies requirements for a quality management system where an organization needs to demonstrate its ability to provide medical devices and related services that consistently meet customer and applicable regulatory requirements" [30].
- "ISO 14971:2007 specifies a process for a manufacturer to identify the hazards associated with medical devices, including in vitro diagnostic (IVD) medical devices, to estimate and evaluate the associated risks, to control these risks, and to monitor the effectiveness of the controls. The requirements of ISO 14971:2007 are applicable to all stages of the life-cycle of a medical device" [31].

4.2   Conformity Assessment

Conformity or compliance assessment involves a set of processes that show medical equipment meets the requirements set forth in the standards related to device type. The main forms of conformity assessment are testing, certification, and inspection. In addition to standards and regulations against which products are assessed, regulatory bodies adopt procedures for specific assessment methods as acceptance test procedure, procedure for regular inspections.

The necessary content of a technical documentation for compliance assessment according to the IEC 60601 is as follows:

- A general description of the device/device family, including any variants planned,
- Design and specifications,
- Label and instructions for use,
- Reference to applicable harmonized standards,
- Results of risk analysis,
- Evidence that the essential requirements have been met.

Dialysis machine design and construction, under referential operating conditions, shall provide for protection against electrical hazard, excessive temperature, and intrusion of fire, dust or water in the device housing. Identification requirements define that dialysis machines shall be marked with the name and trademark of manufacturer, serial number, production year, and model or type reference.

Risk analysis is important for deciding which aspects of a medical device's performance are essential. If variation in a performance may result in injury, then it is considered essential performance. Defining essential performance associated with risk management is crucial for compliance with the IEC 60601. Maintaining risk at acceptable levels is responsibility of the alarming functions of medical device, and alarming functions are built upon accuracy and reliability of the key performance parameters measurement.

Within the European Union medical devices conformity assessment is defined with the essential requirements of the Medical Device Directive (MDD) 2007/47/EC amending the MDD 93/42/EEC [32]. The MDD 93/42/EEC has introduced medical device classification system in order to apply gradual and appropriate conformity assessment procedure. According to the MDD 93/42/EEC dialysis machines are devices with measuring function. With respective to additional rules as duration of contact with the body, degree of invasiveness, and local or systemic effect; dialysis machines are classified as Class IIb medical devices [33] and should follow the procedures referred to in Annex IV, V or VI of the MDD 93/42/EEC, for the "aspects of manufacture concerned with the conformity of the products with the metrological requirements" [34].

5 Dialysis Machine as a Measuring Device

At a national level there are national institutes of metrology (NMIs) responsible to establish and maintain reference standards for these metrological requirements, more specific for the following metrology areas: (1) *scientific metrology* through traceability to the International System of Units, or SI; (2) *legal metrology* through regulated measurements and measuring instruments, and (3) *industrial metrology* through confidence in testing and measurement results via certification, standardization, accreditation and calibration. In case of pending national regulations, directive MDD 93/42/EEC, also IEC 60601 and ISO 62353 refer to standardized safety tests and measurement of output parameters of medical devices related directly to patient.

The IEC 60601 addresses risk management stating: "that the technical description provided by the manufacturer shall include characteristics of the ME equipment, including range, accuracy, and precision of the displayed values." Further on it is defined: "that the instructions for use shall identify the parts on which *preventive inspection and maintenance* shall be performed, by service personnel, including the periods to be applied."

Within the scope of patient safety and risk management the *safety performance inspections* of medical devices and corresponding metrological regulations gain importance. The accuracy and reliability of medical measurements are of direct consequences for the health of a patient, and quality assurance of measurement should be ensured by metrological tools, as calibrations, legal metrological inspections and reference measurement methods [35].

The inspection process should be conducted by a medical device inspection laboratory accredited to ISO 17020 with the objective to determine if specified safety requirements and manufacturer specifications are met. The inspection process procedure should be defined with: purpose and scope, frequency, respective standards, required equipment, permissible error range, assessment process description, remedial action, and required documents. The inspection should be documented by an Inspection Report issued by the accredited laboratory and should comprise Work Order, Measurement Report, Calculation of absolute and relative error of device, and Inspection Certificate [36–40].

In the Sect. 3 the key parameters for a patient safety are identified as: dialysate temperature, dialysate concentration and extracorporeal blood pressure. The measurement components for these parameters require periodic inspections to maintain safe and effective hemodialysis systems.

Measurement range for dialysis machine measurements as defined in respective manufacturer technical documentation [15] are:

- Conductivity (10–17) mS/cm
- Temperature (35–39) °C
- Pressure (−300 to 500) mmHg

Measurement requirements define maximum permissible measurement error as defined in respective manufacturer technical documentation [15] are:

- Conductivity ±1.5% (average)
- Temperature ±3 °C (calibration conditions for dialysate flow of 500 mL/min)
- Pressure ±20 mmHg or ±10% of measurement reading, whichever is greater

6 Dialysis Machine Inspection Procedure

Dialysis machine inspection procedure includes (a) visual inspection and (b) verification of measurement error.

Visual inspection, although not defined in the IEC 60601, are important step for the safety inspections. Visual inspection is a simple procedure confirming that medical device is still in compliance with the manufacturer specifications. Visual inspection include:

- Contamination inspection
- Integrity and functionality inspection
- Markings and labelling inspection

Integrity inspection is visual examination of device and its mechanical parts including: connectors, tubes, cables, sensors, etc. The examiner shall look for cracks, obstructions and other damages.

Functionality inspection is testing of all device functions performance and visual examination of their status on device monitor. Functional inspection includes testing of the alarm system. If the device is damaged or if it fails to perform the functions, the inspection shall be temporary interrupted, and device shall be sent for repair. This remedy action shall be taken with the consent of the institution under the inspection. After repair is performed, depending on the result of the repair, inspection is either continued, or if the device is considered defective or malfunctioning, the device shall be taken out of usage and sent for further remedy action.

Device measurement error verification objective is to confirm that the measurement error is smaller than specified maximum permissible error.

Modern instrumentation systems used for hemodialysis device verification are portable and modular. Dialysis meters are composed of the main display module and sensor modules for measuring specific parameter as: conductivity, pH, temperature and pressure (Fig. 4) of the dialysate fluids supplied by hemodialysis delivery systems. These physical values are primary parameters indicating safe and accurate operation of hemodialysis systems.

In case the dialyses meter is accompanied with the specific software application, during the inspection process the meter shall be connected to the computer running licensed software application for data upload and processing.

Verification is performed to establish indication error for the key dialysis parameters, and verification method shall include range, number of required observations, error type to be calculated, and number of points in calibration curve. Example verification method for the key physical quantities:

- Conductivity: single measurement (10–17 mS/mm), determining relative error,
- Temperature: single measurement (35–39 °C), determining absolute error,
- Pressure: six measurements (−300 mmHg, −150 mmHg, 0 mmHg, 150 mmHg, 350 mmHg, 500 mmHg), determining relative error.

Verification process for pressure sensor linearity is in accordance with the measurement experience of collecting measurement data at six points, at equal intervals along the measurement scale.

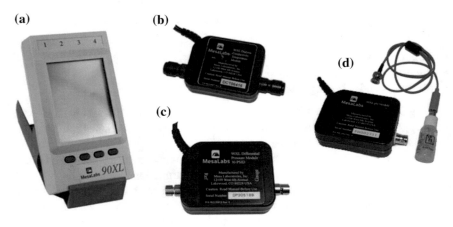

Fig. 4 90XL™ meter (**a**), with conductivity and temperature (**b**), pressure (**c**), and pH module (**d**)

Maximum permissible measurement error pending relevant international regulations, as defined in respective manufacturer technical documentation [15]:

- Conductivity ±1.5% (average)
- Temperature ±3 °C (calibration conditions for dialysate flow of 500 mL/min)
- Pressure ±20 mmHg or ±10% of measurement reading, whichever is greater.

Verification process is described as follows:

Dialysis machine and dialyses meter shall be connected to power supply as marked on the device. If the dialyses meter is battery powered, the batteries should be recharged prior to verification. The sensor modules needed for the verification shall be attach to the meter device with cables that plug into connectors on the meter.

Conductivity and temperature measurements are taken as the dialysis fluid flows through the dialysate circuitry. If performing sample based measurement the sensor end of the conductivity/temperature sensor module is inserted into the container of test solution. Sample solution steadily flow through the sensor module, and when the display is stable, measurement may be taken. For in-line measurement method, conductivity/temperature sensor module is connected directly to the dialysate delivery system. After the flow is re-established through the sensor module measurement may be taken when the display is stable.

Pressure measurement is performed only after one hour post to attaching a pressure sensor module to the meter device. The pressure sensor module may be used for verification of arterial, venous, negative and differential pressure of the delivery system fluid. Intrusion of fluids into the pressure sensor module is prevented with transducer protectors.

Hemodialysis machine verification include inspection of measurement range and permissible measurement error for defined measurement values. If permissible measurement error is larger than stipulated maximum permissible error dialysis machine shall be ordered for service and verification shall be reiterated. Upon completed inspection procedure the inspection officer shall draft inspection results report and mark the device with the appropriate label.

Legal metrology should provide the necessary infrastructure for the traceability of regulated measurements and measuring instruments to SI or national standard, through a documented unbroken chain of calibrations, each contributing to the measurement uncertainty [41]. For medical devices that are applied in the EU but which are not covered by legal metrology in some countries calibration process has to be ensured with an adequate traceability chain.

If medical devices are not covered by legal metrology traceability of results is ensured through calibration procedures with an adequate traceability chain pertaining to the accuracy and correctness of measurement methods. Accordingly, it is necessary to establish adequate traceability chain for measurement results of dialysis machines as medical devices with the measuring function.

For that reason, dialysis meter and verifying sensor modules should be calibrated, according to the ISO 17025 standard, and the results of conductivity, pressure, and temperature measurements can be related to national or international standards, through an unbroken chain of comparisons all having stated uncertainties. Therefore, manufacturers in their specifications and manuals for dialysis meters recommend calibration using traceable reference standards before use or whenever inaccurate readings are suspected [42].

Example of inclusion of dialysis machines in national legal metrology system is Bosnia and Herzegovina. Institute of Metrology of Bosnia and Herzegovina (IMBIH) adopted in 2014 the Measurement and Technical Requirements Rule Book for Dialysis Machines, that establish the metrological characteristics, required of dialysis machines, and which specify methods and equipment for checking their conformity [43].

References

1. Couser WG, Remuzzi G, Mendis S, Tonelli M (2011) The contribution of chronic kidney disease to the global burden of major noncommunicable diseases. Kidney Int 80(12): 1258–1270
2. Graham T (1854) The Bakerian lecture: osmotic force. Philos Trans R Soc London 144, 177–228
3. Graham T (1861) Liquid diffusion applied to analysis. Philos Trans R Soc London 151, 183–224
4. Fick A (1855) Über diffusion. Ann Phys Chem 94, 59–86
5. Ing TS, Rahman MA, Kjellstrand CM (2012) Dialysis: history, development and promise. World Scientific, Singapore
6. Eggerth AH (1921) The preparation and standardization of collodion membranes. J Biol Chem 18:203–221
7. Paskalev DN (2001) Georg Haas (1886–1971): the forgotten hemodialysis pioneer. Dial Transplant 30:828–832
8. Khwaja KO (2006) Dialysis access procedures. In: Humar A, Matas A, Payne W (eds) Atlas of organ transplantation. Springer, London, pp 35–58
9. Kramme R, Hoffmann K, Pozos Azar RS (eds) (2011) Springer handbook of medical technology. Springer-Verlag, Berlin Heidelberg
10. Sharma MK, Wieringa FP, Frijns AJH, Kooman JP (2016) On-line monitoring of electrolytes in hemodialysis: on the road towards individualizing treatment. Expert Rev Med Devices 13 (10):933–943
11. Jungers P, Zingraff J, Man NK, Drueke T (1978) The essentials in hemodialysis: illustrated guide. Springer, Netherlands
12. Gembala M, Kumar S (2011) Sodium and hemodialysis. In: Carpi A (ed) Progress in hemodialysis—from emergent biotechnology to clinical practice. InTech. doi:10.5772/24812
13. Azar AT (ed) (2013) Modeling and control of dialysis systems, vol 1: modeling techniques of hemodialysis systems. SCI 404 Springer-Verlag, Berlin Heidelberg
14. Locatelli F, Covic A, Chazot C et al (2004) Optimal composition of the dialysate, with emphasis on its influence on blood pressure. Nephrol Dial Transplant 19(4):785–796
15. 2008T Hemodialysis machine operator's manual, © Copyright 2000–2015 Fresenius USA, Inc
16. ANSI/AAMI RD52:2004 dialysate for hemodialysis, Arlington, Association for the Advancement of Medical Instrumentation (2004)

17. ANSI/AAMI RD5:2003 hemodialysis systems, Arlington, Association for the Advancement of Medical Instrumentation (2003)
18. Thomas R (2001) Patients' safety and haemodialysis devices. Nephrol Dial Transplant 16 (11):2138–2142
19. Vlchek, DL, Burrows-Hudson S Pressly N (1991) Quality assurance guidelines for hemodialysis devices. U.S. Department of Health and Human Services, Public Health Service, FDA Center for Devices and Radiological Health
20. Lodi CA, Vasta A, Hegbrant MA et al (2010) Multidisciplinary evaluation for severity of hazards applied to hemodialysis devices: an original risk analysis method. Clin J Am Soc Nephrol 5(11):2004–2017
21. International Electrotechnical Commission (2012) IEC 60601-1:2005 +A1: 2012 Medical electrical equipment—part 1: general requirements for basic safety and essential performance, 3rd edn. Geneva
22. International Electrotechnical Commission (2012) IEC 60601-2-16:2012, Medical electrical equipment: part 2–16—particular requirements for basic safety and essential performance of haemodialysis, haemodiafiltration and haemofiltration equipment, 3rd edn. Geneva
23. International Organization for Standardization (2010) ISO 8637:2010 Cardiovascular implants and extracorporeal systems—haemodialysers, haemodiafilters and haemoconcentrators, 3rd edn. Geneva
24. International Organization for Standardization (2010) ISO 8638:2010 Cardiovascular implants and extracorporeal systems—extracorporeal blood circuit for haemodialysers, haemodiafilters and haemofilters, 3rd edn. Geneva
25. International Organization for Standardization (2014) ISO 11663:2014 Quality of dialysis fluid for hemodialysis and related therapies, 2nd edn. Geneva
26. International Organization for Standardization (2014) ISO 13958:2014 Concentrates for hemodialysis and related therapies, 3rd edn. Geneva
27. International Organization for Standardization (2014) ISO 13959:2014 Water for hemodialysis and related therapies, 3rd edn. Geneva
28. International Organization for Standardization (2014) ISO 23500:2014 Guidance for the preparation and quality management of fluids for hemodialysis and related therapies, 2nd edn. Geneva
29. International Organization for Standardization (2014) ISO 26722:2014 Water treatment equipment for hemodialysis applications and related therapies, 2nd edn. Geneva
30. International Organization for Standardization (2016) ISO 13485:2016 Medical devices—quality management systems—requirements for regulatory purposes, 3rd edn. Geneva
31. International Organization for Standardization (2007) ISO 14971:2007 Medical devices—application of risk management to medical devices, Geneva
32. Medical devices directive 2007/47/EC of 5 September 2007, amending council directive 93/42/EEC of 14 June 1993. Off J Eur Union 247 (L), 21–55 (2007)
33. Classification of medical devices. Med Devices 2.4/1 rev. 9, June 2010
34. Medical devices with a measuring function. Med Devices 2.1/5, June 1998
35. do Céu Ferreira M (2011) The role of metrology in the field of medical devices. Int J Metrol Qual Eng 2(2):135–140
36. Gurbeta L, Badnjevic A (2016) Inspection process of medical devices in healthcare institutions: software solution. Health Technol. doi:10.1007/s12553-016-0154-2
37. Gurbeta L, Badnjevic A, Pinjo N, Ljumic F (2015) Software package for tracking status of inspection dates and reports of medical devices in healthcare institutions of bosnia and herzegovina. In: Proceedings of XXV international conference on information, communication and automation technologies (ICAT 2015), Sarajevo, Bosnia and Herzegovina, pp 1–5
38. Gurbeta L, Badnjevic A, Sejdinovic D, Alic B, Abd El-Ilah L, Zunic E (2016) Software solution for tracking inspection processes of medical devices from legal metrology system. In: Proceedings XIV mediterranean conference on medical and biological engineering and computing (MEDICON), 31 March–02 April 2016, Paphos, Cyprus

39. Badnjevic A, Gurbeta L, Boskovic D, Dzemic Z (2015) Medical devices in legal metrology. In: Proceedings of 4th mediterranean conference on embedded computing (MECO), 14–18 June 2015, Budva, Montenegro, pp 365–367
40. Badnjevic A, Gurbeta L, Boskovic D, Dzemic Z (2015) Measurement in medicine—past, present, future. Folia Med Facultatis Med Univ Saraeviensis J 50(1):43–46
41. 90XL™ Meter User's Guide (2014) Mesa Laboratories, Inc. http://dialyguard.mesalabs.com/wp-content/uploads/sites/8/2014/01/90XL-Manual-27AUG2014.pdf. Accessed 23 Aug 2016
42. OIML—International Organization of Legal Metrology (2013) International vocabulary of metrology—basic and general concepts and associated terms (VIM) OIML V 2-200, 3rd edn. 2007 (E/F). https://www.oiml.org/en/files/pdf_v/v002-200-e07.pdf. Accessed 23 Aug 2016
43. Institute of Metrology of Bosnia and Herzegovina IMBIH (2014) Pravilnik o mjeriteljskim i tehnickim zahtjevima za dijalizne uredjaje (Measurement and technical requirements rule book for dialysis machines), Official Gazette Bosnia and Herzegovina, No 75, September 2014. http://sllist.ba/glasnik/2014/broj75/Glasnik075.pdf. Accessed 23 Aug 2016

Inspection and Testing of Pediatric and Neonate Incubators

Lejla Gurbeta, Sebija Izetbegović and Alma Badnjević-Čengić

Abstract Infant incubators revolutionized the way medical care is provided to prematurely born infants or infants born with different kind of diseases and health problems. Due to development of technology infant incubators have been significantly improved from the time they were invented, but the main functions remained the same. Infant incubators provide optimal environmental conditions for infants to recover. This is achieved by controlling parameters such as temperature, relative humidity, air flow rate, oxygen concentration, suitable for medical treatment of certain conditions. Increased sophistication of medical devices, including infant incubators raises numerous questions regarding safety of device and reliability of controlled parameters within medical device. Regulatory bodies define requirements for infant incubators in manufacturing process, distribution and disposal. However, questions on requirements and methods for inspection of safety and performance of these devices are addressed more frequently. These inspections are most usually performed by trained clinical professionals who use specialized devices, analysers (phantoms). Using these calibrated analysers, traceability to international (SI) units of medical device measurement is ensured, which raises reliability of medical device treatment. This chapter describes types of infant incubators that are used in today's healthcare system, as well as requirements on infant incubators stated in international standards and medical device directives. Description of infant incubator analysers is given at the end of this chapter.

L. Gurbeta (✉)
Medical Devices Verification Laboratory, Verlab Ltd Sarajevo, Sarajevo,
Bosnia and Herzegovina
e-mail: gurbeta.lejla@ibu.edu.ba

S. Izetbegović
Clinical Center University of Sarajevo, Sarajevo, Bosnia and Herzegovina

A. Badnjević-Čengić
Canton Hospital Zenica, Zenica, Bosnia and Herzegovina

© Springer Nature Singapore Pte Ltd. 2018
A. Badnjević et al. (eds.), *Inspection of Medical Devices*, Series in Biomedical
Engineering, https://doi.org/10.1007/978-981-10-6650-4_11

List of abbreviations

AAP	American Academy of Paediatrics
AAMI	Association for the Advancement of Medical Instrumentation
AC	Alternating current
ACOG	American College of Obstetricians and Gynaecologists
CM	Corrective maintenance
DC	Direct current
FDA MAUDE	US Food and Drug Administration Manufacturer and User Facility Device Experience Database
IEC	International Electrochemical Commission
ISO	International Organization for Standardization
NICU	Neonatal intensive care units
PM	Preventive maintenance
SI	International System of Units
UNESCO	United Nations Educational, Scientific, and Cultural Organization
UNICEF	United Nations International Children's Emergency Fund
UPS	Uninterruptable power supply
UV	Ultraviolet
WHO	World Health Organization

1 History of Neonatal and Paediatric (Infant) Incubators

Before the Industrial Revolution, prematurely born children and children with disabilities at birth were treated at home with or without the assistance of a doctor. The mortality rate in these cases was extremely high. Development of paediatric and neonate (infant) incubators was the first significant change in treatment of these patients during the 19th century. Stephane Tarnier, the French obstetrician, is considered to be the inventor of this medical device [1]. In first prototype, shown in Fig. 1, installed in a Paris maternity ward, infants were heated up by external hot water reservoir attached to the external heating source. This prototype could house more than one patient and was, later, simplified to a single infant model heated by hot-water bottles replaced manually by the nurse every 3 h. Ventilation relied on simple convection, with air entering at the base and circulating upward around the infant [1]. Since then significant progress has been made in the design, production and application of infant incubators introducing control of other parameters such as humidity, which have been proved to have great impact on patient health during treatment [2].

The first design of infant incubator, with all function as known today, was a work of German paediatrician, Dr. Couney [3]. This prototype, shown in Fig. 2, was more than 1.5 m tall, made of steel and glass, and stood on legs. Water boiler was located on the outside and supplied hot water to a pipe running underneath a mattress made of fine mesh. Thermostat regulated the temperature and air circulation was supported. A pipe carried fresh air from outside the building into the incubator, first passing through absorbent wool suspended in antiseptic or medicated water, then

Fig. 1 Tarnier's incubators in the Paris maternity ward, 1884 [1]

Fig. 2 Courney's incubator [3]

through dry wool to filter out impurities. On top, a chimney-like device with a revolving fan blew the exhausted air upwards and out of the incubator [3].

These devices are currently high-technology electrical medical devices, and they are essential part of care for premature infants and new-borns with health complications (Fig. 3). According to reports from UNICEF and UNESCO in 2000, around 20.6 million infants around the world were born with too little body mass and hospitalization included the use of incubators [4]. Numerous studies conducted until today show that the usage of infant incubators drastically reduced mortality rate, from 66% in 1900 to 38% in the 2000s [5–7].

Dictionary of medical terms define infant incubator as a device for maintaining optimal environmental conditions such as temperature, relative humidity, air flow rate, oxygen concentration, suitable for medical treatment of certain conditions [9], so it can be said that the most important feature of the incubator is its controlled environment. Given that in new-borns, auditory, visual and central nervous system

Fig. 3 Modern intensive care units (ICU) with incubators [8]

have not yet reached a stage of maturity incubator represents an isolated environ-ment in which it is necessary to ensure that the environmental conditions such as noise, light levels, air flow, the value of the temperature does not reach their limits, maximum and minimum, so as not to cause damage to health of the patient.

All medical diagnostic and treatment procedures and devices are designed with purpose and have associated risks that could cause problems in specific circum-stances. Many medical device problems cannot be detected until extensive market experience is gained, even though extensive risk analysis is conducted during the production stage. Safety of medical devices can only be considered in relatively in terms of clinical effectiveness and performance. Clinical effectiveness is a good indicator of device performance, and device performance includes technical func-tions in addition to clinical effectiveness. For example, an alarm feature may not directly contribute to clinical effectiveness but would serve safety purposes. So in order to ensure clinical effectiveness, device performance must be accurate and in accordance with manufacturers specification. Periodic inspection and calibration is the key to ensure device performance is on desired level and to increase the effi-ciency and lifetime of electrical devices including medical devices. Unlike for other electrical devices commonly used, such as electricity meters, scales, various labo-ratory equipment, gas monitors, current management of medical devices, including infant incubators, in healthcare institutions is provided without periodic inspection of performance capabilities, calibrations, and traceability chain. In case of infant incubator, inspection has revealed variation in performance due to multiple factors, including technology, users, environmental conditions, inspection periods, and calibration methods. This represents huge liability in usage of these devices in everyday medical practice since this performance deterioration affect the effective-ness of patient treatment and also lead to serious, life-threatening injuries and death.

2 Paediatric and Neonate (Infant) Incubator Construction and Functionality

Paediatric and neonate incubators today are more sophisticated than the first pro-totypes but basic functions remained the same. They provide controlled environ-mental conditions needed to treat prematurely born infants who are not able to endure all the conditions outside the womb or infants born with certain diseases or health conditions [9].

Patients in incubator have difficulties in thermal regulation because of the poor thermal insulation, relatively large surface area, a small amount of mass to act as a heat sink and they have no ability to conserve heat by changing body position [10]. As shown in Fig. 4, there are four different mechanism of heat lost for patients in incubator that must be taken into account when designing these medical devices. These include radiation, conduction, convection and evaporation. Heat loss through radiation is related to the temperature of the surfaces surrounding the infant but not in direct contact with the infant. Conduction occurs through direct contact with a surface with a different temperature. Heat in incubator is transferred by convection when air currents carry heat away from the body surface and evaporation occurs when water is lost from the skin [11].

Incubators, today, differ in design and accessories depending on manufacturer, but main parts of these electrical medical devices, based on various manufacturer specification, are:

- transparent chamber,
- power supply,
- AC powered heater,
- electric fan motor to circulate the heated air,
- water tank for controlling the humidity of air,
- sensors for measuring temperature, relative humidity, sensor of skin temperature, air flow sensor,
- microprocessor-based temperature controller,
- sir filters,

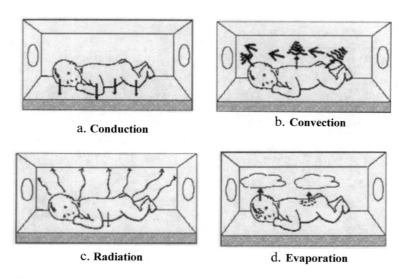

a. Conduction   b. Convection

c. Radiation   d. Evaporation

Fig. 4 Ways of heat loss of the patient in an incubator [11]

- oxygen supply,
- alarms and safety features,
- access ports for nursing care.

Controlled environmental conditions in incubator are achieved by regulation of temperature, humidity, oxygen lever, air flow and sound level. There are two main control modes used in these devices: (1) air temperature control mode and (2) skin temperature control mode.

Incubators can be divided into two groups: (1) stationary and (2) transport incubators, shown in Fig. 5. Difference between transport and stationary incubators are mainly concerned about power supply. Requirements for battery for transport incubators are more complex than for the stationary incubators. When designing transport incubators back up power sources need to be ensured, so if external AC is not available that incubator can switch to external DC source and if that is not available that incubator can use external batteries as power supply to provide stable conditions for transport of patient. Also, most of transport incubators only have air temperature control mode and are supported with lifesaving equipment.

The incubator consist of rigid transparent box, called chamber, which has side openings. Transparent material is used for chamber which need to be designed to resist the strains of transportation and to absorb impacts and ensure thermal insulation properties. Most usually chamber is placed on mount which has ability adjust the angle position of the box. The side openings on the box provide accessibility to patient for resuscitation or procedures without jeopardizing thermal stability, Fig. 5. Most usually in incubator design, openings on the incubator chamber is provided along one of the longer sides of the incubator and provision is made for convenient displacement of the mattress or other infant support laterally out of and back into

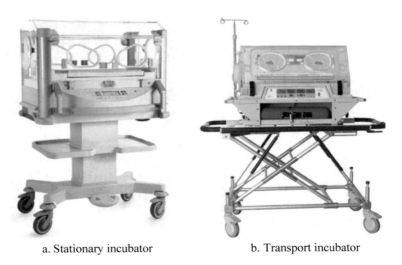

a. Stationary incubator   b. Transport incubator

Fig. 5 Types of pediatric and neonate incubators [12, 13]

the enclosed space of the incubator through the door. At the same time armholes or similar access ports can also be provided, to enable attendants to give attention to the infant within the incubator, without withdrawing the infant support. Longer sides of incubators are intended to be accessible and the one with control panel unit is considered to be front of incubator.

Controlled environmental conditions in incubator mean that the air inside the chamber is held on certain temperature with certain percentage of humidity present in the air. Air intake is usually found at the bottom of the device. The fan with motor drive, placed underneath the chamber, takes the room air and blows it over or through the heating element and the humidifier.

A heating element made from coiled resistance wire known as the tube type (flat or coiled) heater is used in most incubators. To prevent system from overheating, the power rating of heater used in incubator is much less than in the other devices. Typical power rating of heater in incubators is between 100 and 300 W. The heater is controlled by an electronic temperature control unit. The incubator reading, i.e. reading from temperature sensors and reading of relative humidity of air inside of chamber can displayed digitally on control panel of incubator, or using analog indicators.

Humidifier is most usually a water tank. To ensure moisturizing, the heated air flows over the water in the water container and gets moistened, Fig. 6. The humidity is often regulated by closing and opening of valve over the water container closing and opening a deflector plate over the container. Air inlets allow heated and humidified air to enter the chamber, while air outlets take the air from

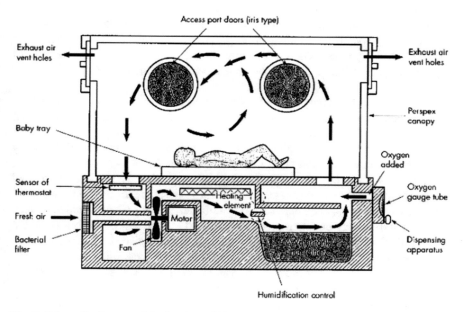

Fig. 6 Schematic diagram of the incubator [14]

the chamber allowing the air circulation inside the chamber. The humidifier should be filled up only with distilled water only in order to avoid corrosive damage to the incubator. The general requirements for incubators in order maintain the body temperature of the baby between 36 and 37.2 °C, are to be able to create an ambient air between 34 and 38 °C with a humidity of 40–90%.

For oxygenation therapy, incubators can be equipped with modules for dosing oxygen into the heated and moistened air. The oxygen concentration is also electronically controlled using microprocessor. If incubator does not poses these module for dosing the oxygen into the air inside the chamber, oxygen can be applied in chamber through a hose connection from an external cylinder, oxygen concentrator or from the central gas supply. Most oxygenation equipment is compatible with most incubators. Alternatively the baby gets the additional oxygen directly through a nasal cannula which pipes oxygen directly into the nostrils.

Maintaining maximum purity of the air circulated in the incubator is enabled by usage of different filters. Also, the purity of air in the incubator can be achieved by introduction of oxygen into the circulation system in a manner to pass through the filter so that not only the replacement air but also the oxygen, when used, is subject to the filtering action. In this way bacteriological impurities originating even in the oxygen supply are substantially eliminated.

Inside the chamber, above the heater and humidifier, cradle for patient is installed. This cradle is most usually equipped with additional scale for measuring the weight of a new-born and can move right, left, up, down. Monitoring and observation equipment is often built into the infant incubator unit which include cardiac monitors, brain-scan equipment, blood-monitoring equipment, thermometers and other instruments for observing vital signs.

For treatment of certain medical conditions, medical professionals use UV (ultraviolet) lamps. These lamps can be put over the incubator chamber since the chamber is transparent. When undergoing treatment with UV light, precautions should always be taken to prevent the unnecessary exposure of UV light to healthy skin.

In order to ensure safety of the infants, all electronically controlled incubators are equipped with alarming system which tracks the state of the environment within the incubator and alarm the medical staff in case of odd settings.

To ensure proper functioning of infant incubators, stable and high-quality delivery of energy is needed to ensure the security of the modern clinical settings. When it comes to an electric failure in the health institution the incubator's internal energy source needs to be able to keep the unit in active mode for few hours in case of power loss. Battery capacity depends on manufacturer. In the case of sudden voltage jumps, operation may be interrupted or may result in incorrect operation. For these reasons, units of uninterruptable power supply (UPS) are placed with incubators.

Beside, infant incubators with closed chamber and controlled environmental conditions that can be stationary or transport, one additional type can be recognized. It is called infant warmer. This medical device consists of a biocompatible bed on which medical professionals place infant and transparent side panels, Fig. 7. The side panels need to fold down for easy access to the patient and can be removed for cleaning. The differences between infant incubators and radiant warmers are given in Table 1.

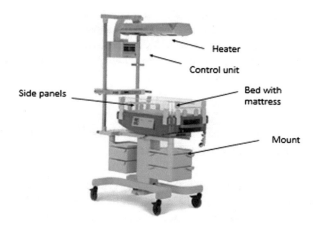

Fig. 7 Infant warmer bed [15]

Table 1 Difference between infant incubators and radiant warmers

Type of device	Incubator	Radiant warmer
Chamber	Closed care	Open care
Heating	Convection principle	Radiation principle
Parameters	Relative humidity control	No relative humidity control
	O_2 control	No O_2 control
Environmental impact	Less disturbances on patient	More external disturbances on patient

The bed of infant warmer is placed on the rail mounting system. Heater is placed above the bed and delivers radiant heat to the patient inside providing unified heat. Heaters in infant warmers are typically made from quartz or ceramic. When designing this medical device, type of heater is very important because it has direct impact on time needed for warmer to meet the desired temperature. Each additional minute of cold stress can lead to increased morbidity for an infant. They are characterized by lower power consumption and long life of the heating element lead to considerable cost savings in healthcare institutions.

These devices also have air temperature control mode and skin temperature control mode. Skin temperature probe monitors infant temperature. The control unit contains the electronic circuits and controls radiant heater and the observation light. The manual mode of operation (air temperature control mode) allows selecting the level of radiant heat output which is indicated by the percentage power displayed on the control panel. The control circuit maintains the selected level of radiant heat. The manual mode has a preheat setting which allows the warmer to be preheated. In the automatic mode (skin temperature control mode) of operation, the patient's control temperature is chosen. A skin temperature probe is used to monitor the patient skin temperature. The control system modulates the radiant heat to maintain the patient at the selected control temperature. In most cases, the patient temperature, control temperature and elapsed time displays are digital for ease of viewing. Visual and audio alarms are present for safety. Difference between infant incubator and radiant warmer is presented in Table 1.

3 Potential Safety Hazards in Therapy with Paediatric and Neonate (Infant) Incubators

Safe and stable environment is necessary in care of new-born infants with certain illnesses or disability that makes them especially vulnerable for the first several months of life, or prematurely born infants. Besides prematurity and low birth-weight, perinatal asphyxia, major birth defects, sepsis, neonatal jaundice, and infant respiratory distress syndrome due to immaturity of the lungs are some of the common diseases that include usage of infant incubator in providing healthcare.

Parameters in the chamber that ensure controlled environment must be regulated as precisely as possible since variations in temperature, humidity, oxygen concentration can have great impact on patient treatment. Table 2 shows the possible complications that can occur, if regulation process of controlling variables is faulty, i.e. situation when temperature, relative humidity, airflow velocity, volume, oxygen saturation and intensity of UV radiation exceed the minimum or maximum values.

The infant's body responds differently to hot and cold temperatures. In the case of hotter environmental temperatures, the infant's body produces, basal metabolic rate increases, causing the body temperature to rise. The risks of hyperthermia are great and should be attended to immediately. Serious overheating can cause heat-stroke or death, and lesser degrees of stress can cause cerebral damage due to hypernatremia dehydration. Babies born more than 8 weeks before term have virtually no ability to sweat. Even in a baby born only 3 weeks early, sweating is severely limited and confined to the head and face. Sweat production matures relatively quickly in the pre-term baby after delivery, allowing the baby to be placed in a regular crib. In the case of cold environmental temperatures, the infant may produce heat by shivering and other muscular activity.

Fan failures (air flow), which are defined as the most critical interferences can cause choking. In addition, incorrect measurement of the level of air flow directly affect oxygen saturation, relative humidity and ambient temperature in the incubator. Several complications like retinal degeneration and dehydration can occur because of errors in the controlling process of UV radiation for the treatment of infants with congenital jaundice (phototherapy), which indicates the importance of reliable and accurate measurement of UV radiation in the incubator.

The level of oxygen saturation and carbon dioxide are the values that are indirectly associated with temperature, relative humidity and air flow from the

Table 2 Possible complications due to usage of faulty infant incubators [16]

Measurement parameters						
Possible complications	Temperature	Relative humidity	Oxygen concentration	Noise level	UV radiation	Airflow
	Sweating hallucinations	Dry skin	Hypoxia	Attention deficit disorder	Bronze child syndrome	Choking
	Weakness vomit	Respiratory diseases	Anemia	Increased blood pressure	Diarrhea	
	Fatigue thirst		Hyperoxia	Expeditious ECG rhythm	Dehydration	
	Headache stroke		Cerebral palsy	Headaches	The degeneration of the retina	
	Muscle cramps		Respiratory diseases	Damage of the inner ear		
	Kidney failure					

environment. Inadequate control of the amount of oxygen supplied into the chamber of the incubator can cause hypoxia or hyperoxia.

Noise level is an important parameter as temperature and humidity is. Over the years it has been proved that sound levels in incubators must be tested and kept within the limits prescribed by standards given that in the case of labour, if the incubator produces too much noise patient can reach permanent hearing loss, Table 3.

An infant incubators are common to paediatric hospitals, birthing centres and neonatal intensive care units (NICU). Comprehensive set of recommendations and standards regarding planning of NICU exist in order to reduce risk of therapy and ensure that all safety aspects are covered. American Academy of Paediatrics (AAP) and American College of Obstetricians and Gynaecologists (ACOG) published several editions of guidelines for construction of NICUs [18, 19] based on clinical experience and an evolving scientific database. These recommendations define unit configurations, location within healthcare institution, clearance and privacy of space, staff requirements, etc. As for European market, no specific NICU standard or recommendation exist until this date. Recommendations refer generally to planning and building as well as management of intensive care units [19]. World

Table 3 The intensity of sound in the incubator [17]

Level	The intensity [dB]	Example	The equivalent within the incubator	Effect
Barely audible	10	Heart rate		
Very quiet	23–30	Silent testimony		<35 dB advisable to sleep
Quietly	40	Standard noises in the house		
	50	Traffic in the distance	Sounds in the background	<50 dB desirable for normal functioning
Gradually hammered	60	Normal conversation	Engine—switching on and off	
	70	Vacuum cleaner	Fan	
Loudly	80	Sounds of traffic or the phone ringing	Hitting fingers in the chamber	
	90	Pneumatic drill	Closing metal door underlying chamber	Possible hearing loss
Very loudly	100	Lawnmower	Closing/opening the door of the chamber	
Uncomfortably loud	120	The sound system in the car	Hitting the head of the mattress	Pain
	140	Plane taking off 30 m away		

Health Organization (WHO) publishes only recommendations on new-born health explaining what health interventions should the pregnant woman, mother, new-born should receive without remark to infant incubators and their placement and maintenance in NICUs. WHO is more concerned on delivering infant incubators and healthcare to remote areas mostly in Africa and Asia to decrease the morbidity rate among premature new-borns in these areas [4]. Beside this WHO issues several guidelines for medical device management in general.

However, despite the existence of recommendations for planning and development of NICUs regarding safety aspects, and potential benefits of infant incubators, the negligent to prescribed safety rules and neglecting use of these devices can potentially lead to severe injuries or death of the patient. Even slight variations in the output parameters, such as a deviation of ± 2 °C in the temperature control in infant incubators, have been proven to have a considerable impact on the survival rate of the patients. Moreover, the correct use of these medical equipment's, especially the incubator, is crucial to maximize the benefits and minimize the risks since their configuration must be targeted to the specific conditions of the patient. Situations where malfunctions of infant incubators lead to serious injuries or deaths are unfortunately not rare. A search on US Food and Drug Administration Manufacturer and User Facility Device Experience database (FDA MAUDE) returns 42 malfunctions, 6 death events and 23 injuries related to neonatal incubators in the last 10 years. Device from various manufacturers caused malfunctions that vary from probe malfunction, alarm system malfunction, to overheating and major power supply malfunctions that were threat to all department where incubator was installed. Over the years using incubators, some of the most commonly observed problems are [19–23]:

- death or injury caused by failure of the thermostat with this leading to hyperthermia patients,
- failures of skin temperature sensor which caused fake display of low skin temperatures and led to overheating of the air in the chamber that is causing hyperthermia patient,
- electrical defects of mechanical components that cause changes in the output values of the parameters in relation to the specification,
- electrical defects that lead to electric shock,
- inadequate regulation of oxygen in the chamber,
- too much noise in the normal operation of the incubator due to failure of mechanical parts,
- Incorrect and incomplete calibration and effect of the human factor in this process.

Diversity and innovativeness of medical devices, significantly contribute to improvement in quality and efficiency of healthcare services. With increasing level of sophistication of medical devices, including infant incubators, reliability and accurate operation is becoming an important safety term that affects quality of healthcare and has potential life risks for patients and medical staff [24]. Basing on experience and evidence so far, these malfunctions seem to be unavoidable even though infant incubators are subject to risk analysis during development and manufacturing process. Risk analysis is conducted according to ISO 14791 Medical devices—Application of risk management to medical devices [25]. This standard specifies a process for a manufacturer to identify the hazards associated with medical devices, estimate and evaluate the associated risks, and to control these risks, as well as to monitor the effectiveness of the controls. In 2012, this standard was harmonized with three European Directives associated with medical devices: (1) Medical Devices Directive 93/42/EEC [26], (2) In vitro Diagnostic Medical Device Directive 98/79/EC [27], and (3) Active Implantable Medical Device Directive 90/385/EEC [28]. This version of standard applies only to manufacturers with devices intended for the European market. The requirements of ISO 14971:2007 are applicable to all stages of the life-cycle of a medical device.

Medical device maintenance, safety and performance inspections are key factors in preventing malfunctions from happening in regular usage of medical devices, including infant incubators. Inspection procedures for determining whether a device complies with specified requirements do not include parts replacement, or device repair, but only determines if specified safety requirements and manufacturer specifications are met. The performance inspection processes are often referred as functional checks. The performance inspections process and electrical safety inspections should be performed periodically as part of preventive maintenance (PM) and always after any repair or C (CM) to ensure that performance, accuracy, safety and performance of device are not downgraded. According to the WHO, PM comprehends all scheduled activities oriented towards extending the life of the device and preventing failure, and can be usually combined with inspection procedures that ensure the correct functionality of the device at the time of the inspections. [4] On the other hand, CM comprehends unscheduled procedures performed based on demand, with the aim to restore the functionality of a malfunctioning device. According to the WHO infant incubators should be revised every 4 or 6 months [4]. Hardware components such as air filter at the incubator, in many cases it is necessary to change every 3 months, or earlier if they notice visible dirt and damage.

4 Standards and Regulations for Paediatric and Neonate (Infant) Incubators

Safety and performance inspection of medical devices is defined by international standards issued by prominent worldwide known organizations such as International Organization for Standardization (ISO) [25], International Electrochemical Commission (IEC) [29], Association for the Advancement of Medical Instrumentation (AAMI) [30], and other national or international regulatory bodies, as shown in Fig. 8, such as Medical Device Agency (MDA) (UK),

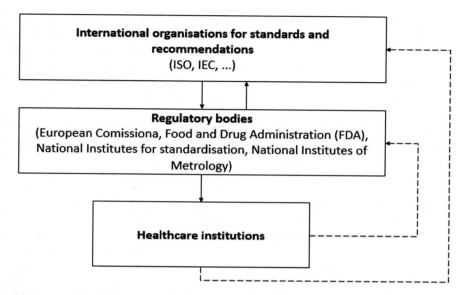

Fig. 8 Regulatory bodies for issuing standards and recommendations on medical devices

Standards Australia (AS)/NZ (Australia/New Zealand), National Fire Protection Association (NFPA)/Association for the Advancement of Medical Instrumentation (AAMI) (USA).

In countries whose regulatory bodies haven't defines national guidance or recommendation for inspection of medical devices, healthcare institutions follow the manufacturer's recommendations. Also, in some of these countries, recommendations for safety and performance inspections have been introduced in legal metrology framework. This has been done assuming that measurements of medical devices are related with S.I. (International System of Units) quantities, such as mass, temperature, length, etc. [31–34]. In this way medical device is recognized as a measuring device that comply with periodical inspections.

IEC 60601 Medical Electrical Equipment—General Requirements for Safety [35] is widely accepted for medical electrical devices and compliance with this standard is requirement for the commercialization of electrical medical equipment in many countries. This standard covers the safety of the design and manufacture of medical electrical equipment. During manufacturing process medical devices are tested for compliance according to this standard to ensure that the design of that equipment is intrinsically safe. The IEC 60601-1 Medical electrical equipment: General requirements for basic safety and essential performance specifies the type-testing requirements for protection against potential electric hazards including protective grounding, ground and patient leakage currents, as well as patient auxiliary currents. The requirements of IEC 60601-1 are applicable to devices that are in usage for a while for periodic inspections and inspections after corrective actions, however, some unforeseen difficulties were presented in practice over a period of time. The main obstacle in applying IEC 60601 for inspections after preventive and corrective actions was ensuring laboratory conditions prescribed by this standard. Laboratory conditions for inspection devices that are already in use are often not applicable, so to overcome this difficulty IEC issued another standard with recommendations for inspection of devices that were already in use, IEC 62353, Medical Electrical Equipment—Recurrent Test and Test After Repair of Medical Electrical Equipment [36].

IEC 62353, defines the requirements of ensuring the electrical safety for medical electronic devices used in the treatment, care, and diagnosis of patients. The standard recognizes that the laboratory conditions described in IEC 60601-1 cannot always be guaranteed when in-service testing of medical devices is undertaken. As a result, test measurements that require certain environmental conditions may not always be applicable or consistent for the testing of equipment that is already in use. Differences between these standards are presented in Table 4.

To incorporate international standards requirements in inspections procedures of in-service devices, many health care institutions use this standard for regular safety and performance testing of medical devices. There is national deviation of this series of standards which include country specific requirements [38], for example EN 60601(EC), UL2601-1 (USA), CSA C22.2 (Canada) and AS/NZ 3200-1 (Australia/New Zealand) [39].

The parameters of the incubator which regularly checking is necessary to make, are part of the multi-dependent variables characterized by nonlinearities and interactions among variables. Temperature and humidity are controlled by the level of electrical current flowing through the two resistors in electrical circuits of these medical devices. Due to various conditions that can evolve if temperature or humidity exceeds certain limits, and deterioration of characteristics of resistors and other elements of electrical circuit these parameters should be subject of periodical inspections. In addition to temperature, there are other quantitative values needed to be controlled such as sound level, airflow, oxygen saturation, and the intensity of ultraviolet (UV) radiation. The parameters of the level of oxygen saturation and UV light intensity are very often not integrated measured in incubators for all existing

Table 4 Differences between IEC 60601 and IEC 62353 [37]

	IEC 60601	IEC 62353
When	During production of device mainly	Before initial start-up, after repair, and periodically
Condition	Laboratory	In service
Earth bond testing	A minimum test current of 25 A	A minimum test current of 200 mA
Under insulation resistance testing	Non existing	Provides three different methods for testing the insulation of medical devices
Leakage testing	Direct method	Three methods that can be used to determine these types of leakage current
Visual inspection	Not clearly defined	Defined

devices. Incubators that have oxygen supplies must have controllers that manage the level of oxygen in the chamber.

The particular requirements and essential performance of infant incubators are defined by IEC 60601-2-19 Medical Electrical Equipment—Part 2: Particular Requirements for Safety of Baby Incubators [40]. IEC 60601-2-20 Medical Electrical Equipment—Part 2: Particular Requirements for Safety of Transport Incubators [41]. IEC 60601-2-21 Medical Electrical Equipment—Part 2: Particular Requirements for Safety Radiant Warmers [42]. These collateral standards are basis for performance inspection of infant incubators. The measurement parameters must comply with afore mentioned standard. These standards determines the variation range of environmental variables in the incubator, such as temperature, relative humidity, velocity of air flow, noise level, etc., as well as methods for evaluating them.

According to the referred IEC 60601-2-19 Medical Electrical Equipment—Part 2: Particular Requirements for Safety of Baby Incubators, IEC 60601-2-20 Medical Electrical Equipment—Part 2: Particular Requirements for Safety of Transport Incubators, measurement of temperature are performed in five positions within the incubator. The standard defines a measurement plane placed 10 cm above of the incubator mattress, where a set of five measurement points exist, Fig. 9. Five temperature sensors must be positioned at certain points in the chamber of the incubator and the humidity sensor must be positioned at central point [40].

Besides the correct positioning of the sensors, a set of six steps must be followed to test the incubator. The environmental temperature during inspection needs to be between 21 and 26 °C. First measurement point is set to be 11 °C over the environment temperature. The incubator must reach the temperature control at the time instant specified by the manufacturer with a tolerance of up to 20%. The incubator

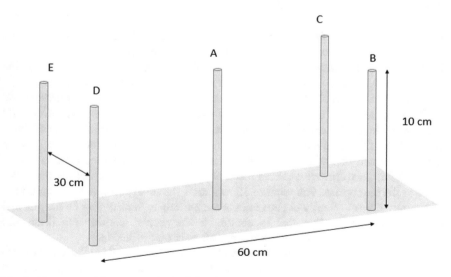

Fig. 9 Placement of five sensors in the infant incubator [40]

temperature control is measured by the incubator sensor. The incubator temperature is adjusted to two operation points, i.e., 32 and 36 °C. At each point, stabilization temperature condition needs to be achieved. This means that the measured incubator temperature at one measuring point must not vary more than 1 °C within the period of one hour. To conclude if incubator complies in respect to temperature requirements, the average temperature measured by the sensors at five points cannot differ by more than 0.8 °C from the average temperature of the incubator within the period of one hour. Moreover, the average temperature cannot differ more than ±1.5 °C from the incubator control temperature. The incubator average temperature is obtained by means of temperatures measured in regular intervals [40]. The temperature distribution requirements should not be met at the expense of high air velocities which will increase evaporative water loss of the patient. The limit 0.35 m/s is derived from measurements on units considered as acceptable in this respect.

Final step in inspection incubators, according to IEC 60601-2-19, IEC 60601-2-20 is verifying alarm system and overshoot temperature. With the incubator temperature at 32 °C, set the temperature control reference in steady state to 36 °C and verify if the overshoot is below 2 °C and that the setting is reached in less than 15 min. The humidity value shown by the incubator sensor must not differ by more than 10% of the value measured by the sensor at central position during the whole period of the incubator functioning. The aforementioned procedures must be performed periodically after each maintenance procedure or when the incubator sensors are found to be uncalibrated.

In normal use the sound level within the compartment shall not exceed sound pressure level of 60 dB. Or this test the infant transport incubator is operated at a control temperature of 36 °C and at a maximum humidity. The background sound level measured inside the compartment needs to be at least 10 dB below that which is measured during the test [40].

To measure the radiant heat from a radiant warmer, according to IEC 60601-2-21 Medical Electrical Equipment—Part 2: Particular Requirements for Safety of Radiant Warmers, an appropriate sensor is being used to capture accurate readings. The temperature probes should be placed in five locations, centre of the mattress, and centre of each quadrant and readings subsequently recorded.

Warm-up time, as shown in Fig. 10, is the time it takes for the incubator to rise 11 °C from ambient temperature. Manufacturers specify the warmup time, and inspection should be carried out to ensure the incubator is within the acceptable range of ±20% the specified warm up time. This is very important, as incubators are often turned on right before use. If the manufacturer specifies their incubator takes 15 min to warm-up it means medical professionals will expect the incubator to be at the appropriate temperature 15 min after the incubator has been turned on. Since new-borns and critical care infants are extremely sensitive to environmental changes, it is imperative that incubators be in proper working condition and warmed to a stable temperature when an infant is placed inside.

UV light intensity, if available in incubator, is also defined by these applicable standards. As for safety inspections, for performing performance inspection various

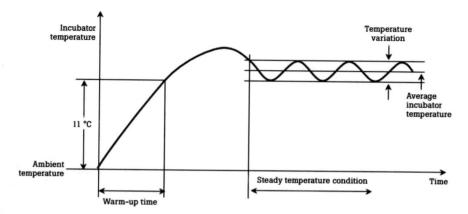

Fig. 10 Inspection of output temperature of infant incubators [40]

analysers available on the market that enable fast and simple automated inspection of performance parameters of infant incubators that will be discussed in next chapter.

Another important aspect for infant incubators is compliance with requirements regarding biocompatibility. International recommendations for biocompatibility inspection can be found in ISO-10993-1:1992, Part I: Biological Evaluation of Medical Devices, Evaluation and Inspection. All parts of the device that have direct or indirect contact with the patient must be checked for potential toxicity that can result from contact of the component materials of the device with the patient's body.

5 Inspection Procedure

Medical device inspection, including infant incubators, consist of safety inspection and performance inspection. Each of these include (1) visual inspection of medical device that is under test and (2) verification of measurement error. This framework is shown in Fig. 11. Safety inspection refers to electrical safety of electrical medical devices used in interaction with patients and performance inspection refers to accuracy of output parameters of medical device. These inspections do not include change of parts or modification of system functionality but they are designed to check if patient relative parameters are in accordance with international require-ments and manufacturer specification.

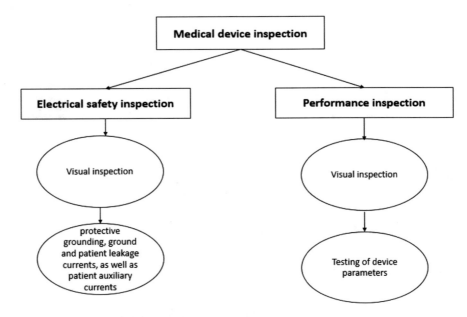

Fig. 11 Framework for medical device inspection

5.1 Visual Inspection

Prior to each functional testing, regardless is it during safety inspection or performance inspection, visual inspection should be conducted. Visual inspection is conducted in order to make sure that the medical equipment in use still conforms to the specifications as released by the manufacturer and has not suffered from any external damage and/or contamination. During the visual inspection, infant incubators is searched for cracks on the incubator chamber, performance of control panel, switches, doors, cables, probes, batteries and other constructional parts. The process of visual inspection is not clearly defined by IEC 60601, however visual inspections form a critical part of the general safety inspections during the functional life of medical equipment. In most cases, 70% of all faults are detected during visual inspection [39].

5.2 Safety Inspection

Every part of electrical medical device that comes into contact with patient's body has some risk of electrical shock caused by unsafe leakage currents. Since neonate and paediatric incubators can be labelled as medical electrical equipment, electrical safety test should be the first test in process of inspection of this medical device.

According to WHO recommendations [34], electrical safety inspection needs to be performed periodically during preventative maintenance, and after every corrective maintenance. They involve testing of ground wire resistance, chassis leakage, patient leakage currents and mains on applied parts.

To perform this inspection, modern technology devices are used. These devices are usually referred as safety analysers and they have predefined tests according to IEC 60601 Medical Electrical Equipment—General Requirements for Safety [35], IEC 62353 Medical Electrical Equipment—Recurrent Test and Test after Repair of Medical Electrical Equipment [36] or other national standards.

5.3 Performance Inspection

Performance inspection is performed in order to determine measurement error of the device under test. This inspection is performed using modern devices that are called analysers or phantoms.

This inspection includes testing of all device functions, including alarm systems. The following ranges of infant incubator parameters should be tested:

- Air temperature: 13–42 °C
- Skin temperature: 13–43 °C
- Relative humidity: 10–99%
- Oxygen: 18–99 vol.%
- Mass: 0–10 kg.

These ranges were stablished by manufacturer's specifications, as well as maximum permissible measurement error pending relevant international regulations:

- Air temperature: ±0.8 °C
- Skin temperature: ±0.3 °C
- Relative humidity: ±10%
- Oxygen: ±5 vol.%
- Mass:

  - ±2% (for mass of 0–2 kg),
  - ±5% (for mass of 2–10 kg).

Performance inspection procedure is defined as follows:

1. Connect the infant incubator and analyzer that is used in procedure to the power supply if needed.
2. Prepare analyzer for measurement (connect all sensors if needed).
3. Place the analyzer in the incubator.
4. Adjust the air temperature in the incubator to the initial measurement temperature;

5. If the incubator has the option of measuring relative humidity and/or oxygen concentration, it is necessary to set the desired humidity and/or oxygen concentration in the incubator.
6. When incubator receahes set temperature wait 5 min and make recording of measured values from all sensors.
7. Set incubator temperature to next measurement point.
8. Repeat procedure until records are made for all measuring points for air temperature parameter.
9. Change to servo control mode (regulated by skin temperature).
10. Set the temperature.
11. Wait until incubator receahes set temperature.
12. Take recordings after 5 min.
     • Repeat procedure for all measuring points. Note: Last measuring point should be at 36 °C to test the overshoot of the device. The temperature should not exceed mentioned permissible error.
13. Verify the alarm systems by exposing device sensors to unusually high and low temperatures, air circulation.
14. When alarm is activated check for sound noise in incubator chamber. It should not exceed 80 dB.

The functional test can lasts an hour and involve measuring air temperature, relative humidity, oxygen concentration at five measuring levels, skin temperature at two measuring levels, and weight at 4 measuring levels and it has been formed according to recommendations defined in IEC 60601. After all measurements are taken, errors are calculated and stated in terms of absolute or relative error depending on the requirements above. After completed inspection procedure the clinical staff, professional performing inspection, should draft inspection results report and mark the device with the appropriate label.

In case that during inspection procedure it has been found that the device under test does not meet requirements it needs to be labelled as faulty and sent to repair. After the device has been repaired, procedure should be repeated.

6   Devices for Inspection of Neonate and Pediatric Incubators

For performing inspection of infant incubators several analysers can be used. For safety inspection there are electrical safety analyser available in the market, Fig. 12, and for performance inspection there are various incubator analysers. These analysers are essential part of inspection protocols which can establish validity of medical devices or systems used in healthcare institutions. In this way comparison between medical devices, and their measurement results, in different healthcare institutions is enabled and overall reliability of medical devices is increased.

a. Fluke Biomedical   b. Fluke Biomedical

Fig. 12 Electrical safety analyzers [43–45]

Electrical safety analysers are design to perform automated inspection of device safety according to applicable electrical safety standards mentioned in previous chapter (IEC 60601, IEC 62353). These analysers are portable and easy to use, and usually they have software support.

For performance inspection, various incubator analysers can be used, Fig. 12. Generally, the incubator analyser is a portable device designed to verify the correct functioning and output parameters of the incubator. All available incubator analysers on the market are mostly units focused on recording the parameters important for the care of infants, such as airflow, sound level, temperature and relative humidity (Fig. 13).

These devices are autonomous data acquisition systems that can be used for measuring and storing the operating parameters of the empty incubator. These devices most usually have data transmission modules via serial communication port, Bluetooth port or wireless port. They have multiple temperature sensors, air flow sensor, humidity sensor and some of them sensor of sound level. Important characteristic of these devices is an internal power supply so that they can be used for monitoring parameters of the incubator in the long run.

During inspection, these devices are placed in an incubator in the same way as a patient, taking into account to provide proper air circulation that the temperature sensors are not blocked and all doors of the chamber of the incubator are closed to ensure adequate temperature measurement. These devices need to be designed to test parameters in accordance with international standards discussed in previous chapter. Inspection incubator by these analysers makes possible to see the current value that incubators exhibit with regard to standards.

Incubator analysers are used to determine whether the output values of the incubator are accurate and reliable or not. According to the results of the analyser, if necessary, the correction is realized by biomedical staff of the medical institution so that the calibration procedure is completed.

Most infant incubators or radiant warmers have an oxygen-controlled environment, as well as contain built-in SpO_2 and monitoring system. These systems are also part of the inspection and preventative maintenance for the incubator/radiant warmer. There are measurement devices for measuring the oxygen saturation and

a. Fluke Biomedical INCU Analyzer I [45]

b. Fluke Biomedical INCU Analyzer II [45]

c. ATOM Medical INCU Tester [46]

d. BIOMED Equipment INCU tester [47]

Fig. 13 Incubator analyzers

Table 5 Minimum specification of incubator analyzer

Parameter	Range	Accuracy
Temperature	20–37 °C	±0.05 °C
Humidity	0–100%	
Sound	10–100 dB	±5 dB(A)
Air flow	0.2–2.0 m/s	±0.1 m/s

the intensity of ultraviolet light, which can be retrofitted or used in devices for periodic inspection of these parameters. The minimum specification of incubator analyser is shown in Table 5.

Both analysers, for safety and performance inspection, are measuring systems that captures multiple parameters at the same time. Not all analysers can be used for inspection purpose. The ranges of measured values and the accuracy of measurement needs to be taken into account. The ranges for inspected parameters such as temperature and humidity are known, so incubator analysers must be able to measure these parameters in range stated in standard with certain accuracy. Chiao

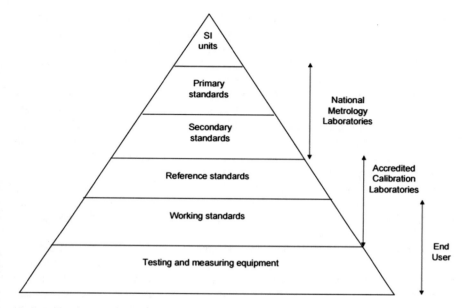

Fig. 14 Traceability chain

et al. [46] discuss on how metrology, traceability and international standards are needed to develop these reference measurements, i.e. phantoms (safety and performance analysers) with stated uncertainty. Uncertainty of these analysers depends on traceability chain, Fig. 14, which is established through calibration.

Each analyser used in inspection process needs to be calibrated in range of values that it measures. To ensure traceability, calibrations need to be performed in laboratories according to ISO 17025 General requirements for the competence of testing and calibration laboratories [25]. The general requirement for choosing analyser for inspection of infant incubator is that it is at least two times more accurate than the device that will be object of the inspection.

The most used method for safety and performance inspection is described in [47], and estimated performance inspection time is about 120 min.

7 Conclusion

Within modern healthcare system there is increased expectation for infant incubators to perform under strict requirements over longer period of time. These devices are just one factor, in very complex system, in treatment of prematurely born or very ill infants, but they role in this process is very significant. Any malfunction of these devices can lead to life threatening injuries or death of patients, which have been reported over period of time. Infant incubators, like other medical devices, have life-expectancy after which performance deteriorates. Managing of medical devices, including infant incubators, is a comprehensive and very demanding task especially in very complex healthcare institutions such as hospitals or clinical centres. Good management strategies should ensure effective, accurate, safe and equal service to all patients.

Medical devices, including infant incubators, are regulated by various international standards and directives issued by national or international regulatory bodies. These standard and directives prescribe all aspects in life cycle of medical device, from manufacturing to disposal including in-service inspections. In many countries there are difficulties in applying these standards and directives in healthcare systems, so assuming measurements from medical devices can be traceable to international standard units (SI units) these devices are introduced into legal metrology system. Nevertheless the method, the primary significance of fulfilling these recommendations in management of medical devices is to ensure safe and reliable medical treatment using these devices.

According to international standards, to ensure infant incubator is reliable for usage, safety and performance inspections should be conducted. These inspections consist of electrical safety inspections and functional testing. Both of these include visual inspection as well. To conduct these inspections various device analysers can be used. These analysers allow technicians to perform automated inspections, with software support, according to international standards in time effective manner. Most usually different analysers are used for inspection of electrical safety and inspection of device performance.

For testing electrical safety in infant incubators, tests according to IEC 60601 or IEC 62353 standard can be used. Even though IEC 62353 complies better with in-service inspection of medical devices, many health care institutions follow the recommendations of IEC 60601. For infant incubators, electrical safety includes testing of grounding, ground and patient leakage currents, as well as patient auxiliary currents.

Performance of infant incubator means inspection of parameters that ensure controlled environment in incubator chamber being air and/or skin temperature, humidity, flow and sound. The inspected ranges and allowed errors of inspected parameters are defined by international standards and medical device directives. For infant incubators, IEC 60601-2-19 (20 and 21) Medical Electrical Equipment—Part 2: Particular Requirements for Safety of Baby Incubators (transport incubators, radiant warmers) are used.

By performing periodical inspections safe and accurate functioning of infant incubators is achieved and more effective healthcare is provided to patients. The purpose of inspection and producing test results is to have a continuous stream of data showing all changes in the performance and safety of the incubators and radiant warmers. Long-term trending of the inspections results provides relevant information for preventive and corrective maintenance. Safety and performance inspections, not only increase reliability of infant incubators, but also lead to optimization of expenses in healthcare institution.

References

1. Baker JP (2000) The incubator and the medical discovery of the premature infant. J Perinatol 20:321. Retrieved 20 Dec 2016
2. Blackfan KD, Yaglou CP (1933) The premature infant: a study of the effects of atmospheric conditions on growth and on development. Am J Dis Child 46(5)
3. Couney MA (1979) Incubator-baby side shows. Pediatrics 64(2):127–141
4. World Health Organisation (WHO). Available at: http://www.who.int/en/
5. Desmond MM (1991) A review of newborn medicine in America: European past and guiding ideology. Am J Perinatol 8:308
6. Cone TE (1983) Perspectives in neonatology. In: Smith GF, Vidyasagar D (eds) Historical review and recent advances in neonatal and perinatal medicine. Mead Johnson Nutritional Division
7. Wertz RW, Wertz DC (1977) Lying-in: a history of childbirth in America. The Free Press, New York
8. Neonatal Intensive Care Unit, picture source: Mayo Clinic. Neonatal Intensive Care. Available at: http://www.mymc.co.il/e/Neonatal_Intensive_Care_Unit/
9. Definition of neonatal and pediatric (infant) incubator. Medical dictionary. Available at: http://medical-dictionary.thefreedictionary.com/incubator
10. Mittal H, Mathew L, Gupta A (2015) Design and development of an infant incubator for controlling multiple parameters. Int J Emerg Trends Electr Electron 11:65–72
11. Soll RF, Allen F (2008) Heat loss prevention in neonates. J Perinatol 28:57–59
12. Types of pediatric and neonate incubators, Stationary incubator. Picture source: http://www.kwipped.com/rentals/medical/infant-incubator/234
13. Types of pediatric and neonate incubators. Transport incubator. Picture source: http://www.medicalexpo.com/medical-manufacturer/neonatal-transport-incubator-30622.html
14. Schematic diagram of the incubator. Picture source: Frank Hospital. Available at: http://www.frankshospitalworkshop.com/equipment/infant_incubators_equipment.html
15. Drager Infant Radiant Warmer. Picture source: Medical Expo. Available at: http://www.medicalexpo.com/prod/draeger/product-68268-421654.html
16. Complications during therapy in infant incubators. Source: Mayo Clinic, Available at: http://www.mayoclinic.org/diseases-conditions/premature-birth/basics/complications/con-20020050
17. Olivera JM, Rocha LA, Rotger VI, Herrera MC (2011) Acoustic pollution in hospital environments. Journal of Physics: Conference Series
18. White RD (2007) Recommended standards for the newborn ICU. J Perinatol 27:S4–S19
19. Collis H (2013) Baby burned to death by faulty incubator at Chinese hospital. Daily Mail UK. Available at: www.dailymail.co.uk/news/article-2368578/Baby-burned-death-faulty-incubator-Chinese-hospital.html. Accessed 25 June 2016
20. US Food and Drug Administration Manufacturer and User Facility Device Experience database, Incubator fault. MAUDE report. Available at: https://www.accessdata.fda.gov/scripts/cdrh/cfdocs/cfmaude/detail.cfm?mdrfoi__id=1905890
21. US Food and Drug Administration Manufacturer and User Facility Device Experience database, Incubator fault. MAUDE report. Available at: http://www.accessdata.fda.gov/scrIpts/cdrh/cfdocs/cfmaude/detail.cfm?mdrfoi__id=1873129
22. US Food and Drug Administration Manufacturer and User Facility Device Experience database, Incubator fault. MAUDE report. Available at: http://www.accessdata.fda.gov/scripts/cdrh/cfdocs/cfMAUDE/detail.cfm?mdrfoi__id=1455792
23. US Food and Drug Administration Manufacturer and User Facility Device Experience database, Incubator fault. MAUDE report. Available at: http://www.accessdata.fda.gov/scripts/cdrh/cfdocs/cfMAUDE/detail.cfm?mdrfoi__id=808764

24. MAUDE—Manufacturer and User Facility Device Experience. Available at: http://www. accessdata.fda.gov/scripts/cdrh/cfdocs/cfMAUDE/Search.cfm?smc=1. Accessed 20 June 2016
25. International organisation for standardisation (ISO). Available at: www.iso.org
26. Directive 2014/32/EU of the European Parliament and of the Council of 26 February 2014 on the harmonization of the laws of the Member States relating to the making available on the market of measuring instruments (recast)
27. https://ec.europa.eu/growth/single-market/european-standards/harmonised-standards/iv-diagnostic-medical-devices/
28. European Commission. Directive 90/385/EEC. Active implantable medical devices. Available at: https://ec.europa.eu/growth/single-market/european-standards/harmonised-standards/implantable-medical-devices_en
29. International Electrotechnical Commission. Available at: http://www.iec.ch/
30. Association for the Advancement of Medical Instrumentation. Available at: https://www. aami.org/
31. Badnjevic A, Gurbeta L, Jimenez ER, Iadanza E (2017) Testing of mechanical ventilators and infant incubators in healthcare institutions. Technol Health Care J 25(2):237–250
32. Gurbeta L, Badnjević A (2016) Inspection process of medical devices in healthcare institutions: software solution. Health Technol J 7(1):109–117 doi:10.1007/s12553-016-0154-2
33. Gurbeta L, Badnjević A, Dzemic Z, Jimenez E.R, Jakupovic A (2016) Testing of therapeutic ultrasound in healthcare institutions in Bosnia and Herzegovina. In: 2nd EAI international conference on future access enablers of ubiquitous and intelligent Infrastructure, Belgrade, Serbia
34. WHO Medical device technical series (2011) Medical equipment maintenance programme overview
35. EC 60601 Medical Electrical Equipment—General Requirements for Safety
36. IEC 62353, Medical Electrical Equipment—Recurrent Test and Test after Repair of Medical Electrical Equipment
37. Comparing IEC 62353 with IEC 60601 for Electromechanical Testing. John Backes. Available at http://www.medicalelectronicsdesign.com/article/comparing-iec-62353-iec-60601-electromechanical-testing
38. National Deviations to IEC 60601-1 by MDDI; National Deviations to IEC 60601-1 by Eisner Safety Consultants
39. A Practical guide to IEC 60601-1-Rigel Medical
40. IEC 60601-2-19: Medical electrical equipment. The particular requirements and essential performance of infant incubators. www.iso.org. Accessed 25 June 2016
41. IEC 60601-2-20: Medical electrical equipment. The particular requirements and essential performance of transport incubators. www.iso.org. Accessed 25 June 2016
42. IEC 60601-2-21: Medical electrical equipment. The particular requirements and essential performance of radiant warmers. www.iso.org. Accessed 25 June 2016
43. Electrical safety analyzer. Picture source available at: http://www.flukebiomedical.com/biomedical/usen/products/electrical-safety-analyzers
44. Electrical safety analyzer. Picture source available at: http://bmet.wikia.com/wiki/Safety_Analyzer
45. Electrical safety analyzer. Picture source available at: http://www.rigelmedical.com/userfiles/safetest-60-electrical-safety-analyzer.php
46. Chiao JC, Goldman JM, Heck DA (2008) Metrology and standards needs for some categories of medical devices. J Res Natl Inst Stand Technol 113(2):121–129

47. Clark T, Lane M, Rafuse L, Medical equipment quality assurance: inspection program development and procedures medical equipment quality assurance. Fluke Biomedical% Vermont University, Revision 2
48. Incubator analyzer. Picture source available at: http://www.flukebiomedical.com/biomedical/usen/products/incubator-analyzers
49. Incubator analyzer. Picture source available at: http://www.atomed.co.jp/english/product/cat_neonatology/detail/151
50. Incubator analyzer. Picture source available at: http://www.aliexpress.com/w/wholesale-incubator-testers.html

Inspection and Testing of Infusion Pumps

Ernesto Iadanza, Diletta Pennati and Fabrizio Dori

Abstract Infusion pumps are ME (Medical Electrical) devices intended to regulate the intravenous flow of liquids through a positive pressure generated by the pump itself. Infusion pumps are used in all those wards where patients need life support and a proper nutrition. Even if infusion pumps became almost totally automated devices, a proper training of users is still necessary. In order to reduce the number of errors reported, the first step to do is to teach to all users how to prevent serious problems to patients. It is necessary to identify what are the potential risks that may be encountered in using these devices and especially the sources from which they derive. For each kind of hazard a protection system must be available. Maintenance tasks and accuracy tests are described in this chapter. Also critical issues in management and safety are described and analysed. The designers of smart infusion pumps must make sure to create products with functions that can be easily understood and used by the operators. Only thanks to the collaboration between manufacturers, technicians, doctors and nurses it becomes possible to organize the maintenance of any device, infusion pumps included.

1 Introduction

1.1 Brief History of Infusion Pumps

Although the first recorded attempt at intravenous medicine dates to 1492, this branch of medical science gained real momentum in the 17th century. The first working IV infusion device was invented by an English architect, Christopher Wren, in 1658. Soon afterward, medical scientists conducted ever increasing experiments with administering drugs and fluids intravenously. However, these early experiments also led to some deaths, so the governments decided to ban the use of infusion devices for hundreds of years. In the 19th century, early prototypes

E. Iadanza (✉) · D. Pennati · F. Dori
Department of Information Engineering, University of Florence, Florence, Italy
e-mail: ernesto.iadanza@unifi.it

© Springer Nature Singapore Pte Ltd. 2018
A. Badnjević et al. (eds.), *Inspection of Medical Devices*, Series in Biomedical
Engineering, https://doi.org/10.1007/978-981-10-6650-4_12

of infusion pumps were invented and the key elements of intravenous transfusion, which are still observed today, were established: a slow infusion process, awareness and prevention of risks from air embolism, and avoiding volume overload. The 20th century saw huge advances in intravenous medicine including IV pumps, mostly during the two World Wars when needles were refined, rubber tubing was replaced by plastic, and vacuum bottles that reduced the risk of air embolism were designed. In the early 1970s, Dean Kamen invented the first wearable infusion pump, which represents the modern ambulatory pump. This advancement was very useful for diabetic patients, because it is able to automatically administer precise doses at regularly timed intervals. At the beginning of the 21st century, smart pumps started to diffuse, which included built-in drug libraries and guidelines, an electronic record of all alerts and the ability to link to a hospital's information system, showing a constant stream of information to medical staff.

Infusion pumps are ME (Medical Electrical) devices intended to regulate the intravenous flow of liquids through a positive pressure generated by the pump itself. Infusion pumps are used in all those wards where patients need life support and a proper nutrition. The intravenous administration allows avoiding the step of the liquid's absorption (drugs or nutrients) that would slow down the efficacy. It was estimated that about 90% of hospitalized patients receive medications through infusion pumps [1].

There are different kinds of infusion pumps, classified according to flow type:

- *Type 1*. Continuous infusion
- *Type 2*. Non-continuous infusion
- *Type 3*. Bolus infusers
- *Type 4*. Combination of different kinds of infusers on the same device (type 1 combined with type 2 and/or 3)
- *Type 5*. Programmable profile pumps.

Furthermore, they are classified according to the infusion type:

- *Volumetric infusion pumps*. The infusion rate is set by the operator in terms of volume per unit of time.
- *Drop infusion pumps*. Unlike the previous ones, the rate is shown in terms of drops per unit of time.
- *Syringe pumps*. Liquids infusion is controlled by one or more single-action syringes. The delivery rate is set in terms of volume per unit of time.
- *Infusion pumps for ambulatory use*. These pumps are designed so as to be continuously carried by the patient.

Since these are devices that affect patient's safety and sometimes a patient's life itself, infusion pumps need an adequate preventive maintenance, both by qualified technicians and by hospital stuff, such as doctors and nurses. Since these devices could also be handled by the patients themselves or in general by not fully trained subjects, the functionality of these devices may be compromised, thus causing possible serious damage to the patient. According to ECRI's report "Top 10 Health

Technology Hazards for 2016", dated November 2015, the hazards related to improper use of infusion devices are listed in the top ten hazards in using health technologies [2]. In particular, in the second place you can find the missed alarm detection from a medical device such as the infusion pump, or the inadequate interpretation of this potentially hazardous situation, which can lead to serious injury or even to death. In third place there are reported risks associated with an excessive amount of drug infused into the patient, due to a wrong programming of the device, in terms of concentration or type of medical product. The potential hazards are not listed according to their frequency or their danger, but according to their priorities, that is how much it is urgent to intervene to prevent them. Hence the emphasis on the importance of prevention is both through careful and controlled technology management and through a correct interpretation of the applicable standards.

In order to guarantee an acceptable safety level, the design and the construction of the infusion systems must meet the requirements from the following European harmonized standard:

- EN 60601-2-24:2015-05 [3], which contains the requirements for basic safety and essential performance of infusion pumps and their relative controllers. The general standard of reference is the IEC 60601-1 [4].
- EN ISO 14971: 2012 [5], which regards to risk analysis associated with medical devices.

Also some national guides and recommendations for infusion systems maintenance can be useful. These documents are addressed both to users and to qualified technicians. In fact, the maintenance of any device is complete when tests and periodic checks are accompanied by inspections before use, which are the first ones that allow identifying any hazards to the patients.

1.2 Motivation for Inspection

Even if infusion pumps became almost totally automated devices, a proper training of users is still necessary. The American agency for Food and Drugs Administration (FDA) received about fifty thousand of reports between 2007 and 2016, in which both common users, such as nurses and patients themselves, and clinical engineers describe adverse situations during the use of an infusion pump [6]. In order to reduce the number of errors reported, the first step to do is to teach to all users how to prevent serious problems to patients. In fact, especially after the smart pumps' advent, users started to rely too much on technological advancements, such as computerized recognition of a patient's identity and his corresponding drug and dose to administrate. Without a responsible behavior, it is impossible to guarantee a safe infusion, for this reason users should know which electrical, mechanical and also software component of the device could be malfunctioning and could lead to improper infusion. Naturally, it is not requested to unauthorized staff to perform any kind of parameters measurements, but only to visually control the integrity of the pump and of its accessories before each use.

2 Types of Hazards Associated with the Use of Infusion Pumps

Before analyzing in detail the check tests carried out by the technical staff, it is necessary to identify what are the potential risks that may be encountered in using these devices and especially the sources from which they derive. According to the abovementioned standard, for each kind of hazard a protection system must be available. Some checks are only visual inspections, while other tests require additional instrumentation provided by the manufacturer.

- *Protection against electrical hazards.* Since infusion pumps are ME devices, they have live parts or that might become live in case of failure. Therefore, an adequate insulation must be guaranteed both for supply cables and for applied parts that, for this reason, must be floating.
- *Protection against mechanical hazards.* Danger of collision, cut, fall, friction, instability and vibrations are part of this category. These hazards could damage the patient both directly and indirectly, since the quantity of transmitted liquid may be altered (especially in infusion pumps for ambulatory use).
- *Protection against supply mains interruption.* The standard says that in devices connected to a supply main, only a low priority alarm signal is provided and it is held for almost 3 min. Instead, in devices with only an internal electrical power source the alarm should start almost 30 min before the battery runs out. If both supplies break down, a high priority alarm must be given. In any way the delivery must be interrupted.
- *Protection against over-infusion.* An alarm signal must be initiated under single fault condition and the infusion must be ceased or reduced to the KOR (Keep Open Rate), so as to maintain the patient line open. As soon as the administration set is installed, verify that there is not an over-infusion resulting from a free flow condition.
- *Maximum infusion pressure.* The ME equipment shall not produce a pressure capable of causing ruptures or leaks in the administration set.
- *Protection against unintended bolus volumes and by occlusion.* Any occlusion of the administration line can lead to insufficient infusion of liquid and, after the alarm activation, to the administration of an unintended bolus.

In Fig. 1, the test apparatus to determine the occlusion alarm threshold and bolus volumes is represented [3].

After filling with a test solution, the administration set and the rigid tubing, connected with the pressure transducer, the intermediate rate and the minimum

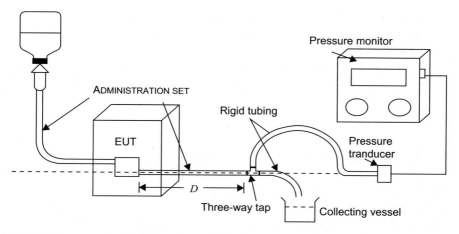

Fig. 1 Test apparatus to determine the occlusion alarm threshold and bolus volumes

occlusion alarm threshold are selected. The end of the patient line is connected to the three-way tap, which is opened to the collecting vessel. After activating the device under testing (EUT), wait for the flow becoming constant. Then the tap is closed again and the time between the closing and the alarm activation is calculated. In order to measure the bolus volume accumulated, the tap is reopened and the liquid is collected in the previously emptied vessel until the pressure returns to the atmospheric level (see on the monitor). The results obtained should be compared with the threshold values specified by the manufacturer.

- *Reverse delivery.* Both in normal use and in single fault condition, any reverse flow shall not cause an unacceptable risk for the patient.
- *ME equipment and drop sensor orientation.* The operation of the apparatus must be safe even when the drop chamber is tilted or it is incorrectly filled. For this reason, a test is carried out in normal conditions, by titling the chamber to a maximum of 20° from the vertical position and visually analyzing the resulting effects. If the liquid infusion is altered, an alarm signal activates and the flow is interrupted.
- *Protection against air infusion.* The device must protect the patient from any air infusion, which could cause injuries, such as formation of emboli. These air infusions can result from incorrect priming of the set, lost connections or administration set running dry. Some devices allow air bubbles to be trapped, which can be removed without the need to disconnect the administration set [7]. However, an alarm signal is generated and it is impossible to recommence liquid delivery by a single action. These requirements are not applied to syringe and ambulatory pumps because it is natural to find small air bubbles dissolved in the liquid into the containers, which are not hazardous for the patient.

- *Protection against under-infusions.* The presence of air inside the tubes or an insufficient pressure of the liquid could make sure that the patient does not receive a sufficient infusion.
- *Protection against leakage of liquids.* Any leakage of liquids from containers, tubing or couplings, or water inputs from outside must not wet uninsulated active components or other critical parts of the device, compromising the safe use.

3 Maintenance by Qualified Technicians

The II level maintenance is periodically (generally once a year) carried out by qualified technicians where the device is used, in order to highlight any failure and/or malfunction. It is important that a check takes place before using the device as soon as it is delivered, possibly in the presence of the manufacturer or its representative. The maintenance consists of a few basic steps:

- *Qualitative trials.* Technicians must verify the devices entireness (packing, supply cables, connectors, casings), the cleanliness of the infusion set and of all the parts accessible from the outside (valves, needles, displays, buttons). Moreover, it is necessary to analyze the whole enclosed documentation and the rating data that characterize the device and facilitate the tracing in the building.
- *Quantitative and instrument trials.* Besides the visual inspection, it is necessary to measure the important parameters for safety purposes, such as leakage currents, insulation resistance, and resistance of the protective earth conductor. Moreover, the operation of protection systems already listed in paragraph 2 and the accuracy of operating data for each pump must be verified. It is essential that all measures are carried out while the machine is not connected to the patient.

In addition to these tests, technicians must make sure that maintenance by operators was successful and they should implement all the maintenance programs prescribed by the manufacturer.

This approach is generally coming from the technical context content in IEC 60601-1 and IEC 60601-2-24 technical standard. Among the various elements characterizing this regulatory environment it is certainly worth to highlight one aspect that perhaps for these devices is the key security element. This element is contained in section 201.12 "Accuracy of controls and instruments and protection against hazardous outputs" of IEC 60601-2-24.

According to general standard "When applicable, the manufacturer shall address in the risk management process the risks associated with accuracy of controls and instruments", where compliance is checked by inspection of the risk management file. In fact, the essential safety component is "the ability of the equipment to maintain the manufacturer's stated accuracy" regardless of clinical criteria of the patient (age, drugs used, etc.). General formula is described in paragraph 201.12.1.101, and compliance is checked, using the tests prescribed in

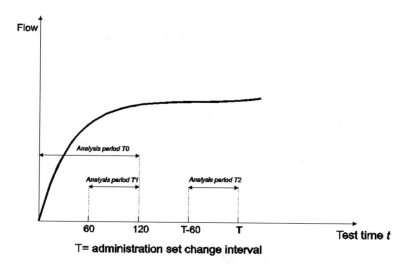

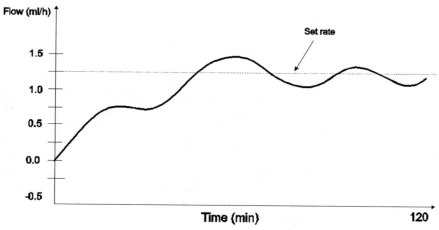

Fig. 2 Accuracy tests for volumetric infusion controllers, volumetric infusion pumps and syringe or container pumps

201.12.1.102–201.12.1.108: the goal is to verify the accuracy of the device according to its defined type and the Manufacturer's disclosure of accuracy. For instance, accuracy tests for volumetric infusion controllers, volumetric infusion pumps and syringe or container pumps are shown in Fig. 2. A test solution of ISO 3696:1987 Class III is needed as well as installing an unused administration set and setting up the equipment with the test solution in accordance with the Manufacturer's instructions for use. After this setting test parameters proceeds with plotting flow Q_i (ml/h), against time T_0 (min) for the first 2 h of the test period. Then plot percentage variation E_p (max.) and E_p (min.) against observation window duration P (min) and the overall mean percentage error A (derived from Eq. 5) measured over the analysis period T_1 (min) of the second hour of the test period. Plot percentage variation E_p (max.) and E_p (min.) against observation window duration P (min) and the overall mean percentage error B (derived from Eq. 6) measured over the analysis period T_2 (min) of the last hour of the test period. Note that E_p (max.) is "maximum measured error in observation window of specified duration" and E_p(min.) is "minimum measured error in observation window of specified duration", calculated using the "trumpet algorithm" as described in the same technical document, while the same document also contains an example of Start-up graph plotted from data gathered during the first 2 h of the test period.

$$Q_i = \frac{60(W_i - W_{i-1})}{Sd} \left(\frac{\mathrm{ml}}{\mathrm{h}}\right) \tag{1}$$

- W_i is the ith mass sample from the analysis period T_0 (g) (corrected for evaporative loss);
- T_0 is the analysis period (min);
- S is the sample interval (min);
- d is the density of water (0.998 g/ml at 20 °C)

Leakage currents check

This type of measurement is essential to guarantee the safety of the patient, since leakage currents might run also through the applied parts, that for this reason should be BF or CF. All these tests must be done under normal condition and single fault condition. Normal conditions are described in paragraph 8.1a) of IEC 60601-1 [4], followed by paragraph 8.1b) where single fault conditions are listed. Regarding infusion pumps, the particular standard IEC 60601-2-24 remembers some specific conditions that must be considered as normal instead of single fault conditions. In fact, according to paragraph 201.4.7 [3]: "The following are not regarded as single fault conditions, but are regarded as normal conditions:

- leakage from the administration set and/or the liquid supply;
- depletion of the internal electrical power source;
- mispositioning and/or incorrect filling of a drop chamber;
- air in the supply line or that part of the medical equipment within which flow regulation, flow shut-off or air detection occurs;
- pulling on the patient line (see ISO 8536-4)."

There are different tests to check leakage currents, using different test devices.

(1) *Measuring circuit for leakage currents through the patient (through applied parts).* During the tests, the apparatus must be insulated from earth and connected to the appropriate measurement device. Two measurements are performed; the second one with a reverse polarity compared to the first one, and the highest value is noted as "first measure" value. In the case of earth leakage

currents, the first measure value must not exceed 0.5 and 1 mA in single fault conditions. Instead, if the leakage current flows through the patient, the maximum values are 0.1 and 0.5 mA for BF applied parts and 0.01 and 0.05 mA for CF applied parts. Anyhow, the following measurement should not be greater than 1.5 times the value of the first measure. In Fig. 3 the measuring circuit for leakage currents through the body (modeled with a NaCl saline solution at 0.9%, called physiological solution) is represented. The patient line is represented by the upper container, while the patient connection between the device under test and the measuring device are immersed in the lower container [3].

(2) *Measuring circuit for leakage currents to earth.* Even if the current doesn't flow through applied parts, it is necessary to protect the patient and eventually the operator from any kind of leakage currents. The test apparatus is shown in Figure 15 of IEC 60601-1 [4], if the leakage current is caused by an internal voltage. In paragraph 8.7.4.7 the standard specifies that during the test "a) …an enclosure, other than an applied part made of insulating material is placed in any position of normal use upon a flat metal surface connected to earth with dimensions at least equal to the plan-projection of the enclosure." Instead if it is caused by an external voltage on a signal input/output part, the measuring circuit to be used is represented in Figure 17 of the same standard.

More allowable values of patient leakage currents under normal condition and single fault condition are listed in Table 1, which has been extracted from IEC 60601-1 [4]. As it is possible to expect, SFC allow more permissive values of leakage currents then in normal conditions but, as specified previously, any values could not exceed 1 mA.

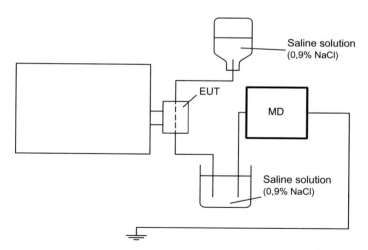

Fig. 3 Test apparatus for leakage currents through the patient

Table 1 Allowable values of leakage currents (values are expressed in µA)

Current	Description	Reference	Measuring circuit		Type bf applied part		Type cf applied part	
					NC	SFC	NC	SFC
Patient leakage current	From patient connection to earth	8.7.4.7 a)	Figure 15	d. c.	10	50	10	50
				a. c.	100	500	10	50
	Caused by an external voltage on a SIP/SOP	8.7.4.7 c)	Figure 17	d. c.	10	50	10	50
				a. c.	100	500	10	50
Total patient leakage current[a]	With the same types of applied part connected together	8.7.4.7 a) and 8.7.4.7 h)	Figures 15 and 20	d. c.	50	100	50	100
				a. c.	500	1000	50	100
	Caused by an external voltage on a SIP/SOP	8.7.4.7 c) and 8.7.4.7 h)	Figures 17 and 20	d. c.	50	100	50	100
				a. c.	500	1000	50	100

Key
NC Normal condition
SFC Single fault condition
[a]Total patient leakage current values are only applicable to equipment having multiple applied parts. See 8.7.4.7 h). The individual applied parts shall comply with the patient leakage current values

Protective earth conductor check
The measure of the resistance of the protective earth conductor is performed between the mains plug protection contact and all metal parts of the apparatus that may come into contact with live parts. The measured value must not exceed 0.2Ω.

Insulation resistance check
The insulation resistance between the mains electricity and ground should not be less than 3 MΩ.

4 Accuracy Tests for Infusion Pumps

A problem associated with the use of infusion pumps regards their use in the first instants of operation. In fact, they will not start immediately to dispense the liquid in accordance with the speed and the mode set by the operator, but there is a settling time in which the device stabilizes at the programmed values. Furthermore, it is possible that during normal operation the device has undesired fluctuations, caused by inadequate administration sets for that kind of pump, unusual characteristics of infusion fluid (generally related to the viscosity and the concentration of the substances in its interior), a use of too thin needles or an inadequate protection against external environmental conditions (such as temperature or vibrations). All of these factors affect the accuracy of the infusion pump and consequently the maintaining of safe performance over time. For this reason, data on performance following the start of infusion are important: graphs which give a good indication of the nature of short-term fluctuations should be included in the instructions for use, in order to allow the operator to determine the start-up performance of the pump and the type of output (continuous, discontinuous, cyclical). To be aware of any possible fluctuations while using the device, the tests must be carried out after the rate of the infusion flow has stabilized, so that the results do not depend solely upon the first minutes of operation.

In the following paragraphs, accuracy tests for each type of infusion pumps are described. When making the tests, it is fixed with a given period of time for measuring the flow and the parameters that must be respected in order to follow a certain standard. Then, the obtained data is compared with those supplied by the manufacturer. All the tests are performed using ISO Class III solution water for medical use. Moreover, the technician that carries out these tests is required to use administration sets, syringes, needles and all other elements of the same type recommended by the manufacturer.

4.1 Volumetric Infusion Pumps

This type of pump is designed to infuse precise volumes of liquid at controllable speed, directly intravenous, using different gauge needles and liquids of all types. They are provided with volumetric infusion controllers, which allow regulation of the liquid delivery to the patient. In particular, they use gravity to supply the required infusion pressure. Generally, the minimum rate is set to 30 ml/h, the intermediate rate to 120 ml/h. After reaching the steady-state condition (within maximum of 5 min), the test is carried out for a period of 120 min at the intermediate rate, with or without creating a back pressure of ±13.33 kPa. In that case, the manufacturer shall disclose the maximum deviation between the result under normal conditions and under back pressure conditions. In Fig. 4 the apparatus for this kind of tests is outlined [3]. As can be seen, the median line of the pumping chamber must be at the same height of the needle tip. The height h_1 between the liquid level and the device undergoing testing shall be such, that the pressure that is generated is greater than the maximum venous pressure, so that the liquid could flow.

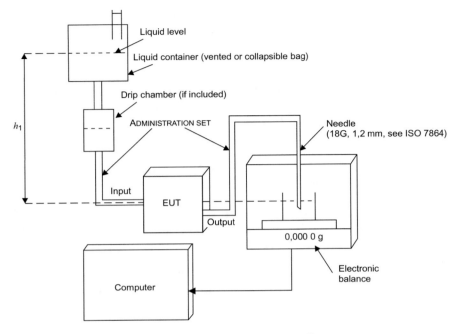

Fig. 4 Test apparatus for volumetric infusion pumps and controllers

4.2 Syringe Infusion Pumps

The type of test is the same performed for the volumetric infusion pumps, but through a different test apparatus, shown in Fig. 5 [3]. In this case, the syringe must be placed at the same level of the needle tip. Also the intermediate rate is lower, selectable to 25 ml/h.

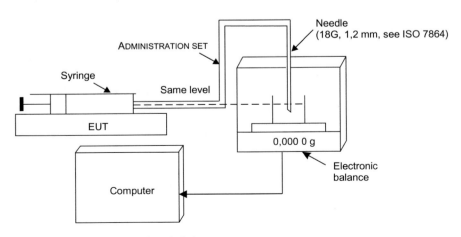

Fig. 5 Test apparatus for syringe infusion pumps

4.3 Drop Infusion Pumps

In these kinds of pumps the rate is adjusted according to the number of drops per minute. Therefore, this value will vary with liquid viscosity, in such a way that the apparatus accuracy is guaranteed for any type of liquid. The intermediate rate is set to 50 drops per min and also in this case the test could be carried out with or without creating a negative back pressure of −13.33 kPa.

As already stated in the introduction, the infusion pumps can also be classified according to the type of flow. If it is a type 1 pump, it is designed to guarantee a continuous flow, the flow pattern over time is monitored at regular predetermined intervals (usually 15 min) and the volume of infused liquid at each sampling interval is measured, until the container is not emptied. If it is a type 2 pump, the flow is not continuous and the sampling interval is chosen as a multiple of the duration of each single infusion. Then, as in the previous case, the volume of infused liquid for each range is measured, covering a period of 100 sampling intervals. If it is a type 3 pump, it is able to administer a bolus of drugs, the test must be repeated for almost 25 times at minimum rate. All the boluses are collected in a single container and the difference between the quantity delivered and that provided by the manufacturer is verified. For infusion pumps of type 4 and 5, the various types of tests are performed according to the programmed operation.

4.4 Graphs

The test protocols are devised to characterize the steady-state flow and to identify errors both in mean and in variation about the mean. A graph of flow versus time gives a clear and simple picture of the general stability with time. In Fig. 6 the so-called "start-up curve" is represented: it can be noted how with the passage of time the flow of the liquid tends to settle on the value set by the operator [3].

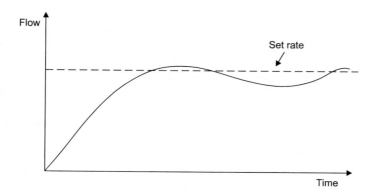

Fig. 6 Start-up graph

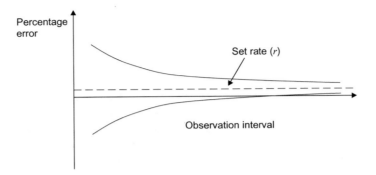

Fig. 7 Trumpet curve

After flow stabilization, a graph could be formulated to quantify the variations in mean flow accuracy, called "trumpet curve", represented in Fig. 7 [3]. An observation interval is fixed, in which the variations are presented as maximum and minimum deviation from the average total flow. In the graph, the two continued curves represent the maximum and minimum measured error, while the broken line represents the desired flow rate, at which the two curves tend. Therefore, this graph allows comparing the performance of the device with that provided by the manufacturer. However, it is possible that the sampled flow data could be susceptible to measurement anomalies: these may be due both to external environmental factors and to the formation of air bubbles, and to errors in the data sampling. All these reduce the quality of the sampled data, then also the reliability and the reproducibility of the obtained curve is decreased. Also for this reason, data starts to be sampled only after the flow has been stabilized on the steady-state value, in order to eliminate anomalies related to the settling period.

4.5   Alarms Analysis

As already affirmed in paragraph 2, whenever a hazardous situation occurs, an alarm, both optical and auditory, is activated. Depending on the severity, and consequently on the timeliness with which an operator must intervene, a priority to each alarm, indicated as high or low, is assigned. Examples of situations characterized by a high priority alarm are ME equipment failures, sudden end of infusion, occlusions, presence of air in line. Instead, events like the premature infusion interruption, the battery depletion, or a stand-by situation of the pump have low priority. Regarding the acoustic alarm volume, it should generate a sound-pressure level of at least 45 dB at 1 m and shall not be adjustable by the operator below this value without the use of particular tools or access codes. This function of regulation is not provided for ambulatory infusion pumps: in fact, since they are transportable and therefore also usable outside the hospital room, it is necessary that the alarm should be audible even in not very quiet and crowded places. Infusions pumps have a function that allows the alarm to be silenced for a determined period, which must not exceed 120 s, beyond which the intervention of an operator becomes necessary.

According to a study published in 2012 in the British Journal of Nursing [7], the ten main causes of alarm activation are shown in the pie chart in Fig. 8. The analysis was carried out on one type of volumetric infusion pump, used in a hospital in the United Kingdom, and it was conducted by the MHRA (Medicines and Healthcare products Regulatory Agency). The occlusion of the administration route is the most frequent cause, as well as the most serious. It is remarkable that the alarm named as "on-hold" is activated after a period of two minutes of inactivity of the pump.

While some potential hazards have already been analyzed in the previous paragraphs, particular attention should be given to battery alarms. It is significant that the number of alarms due to dead batteries is lower than the number of alarms for almost exhausted, but still functional batteries. All this is possible only if there is a good policy of management and maintenance of the equipment, which allows the replacement of the batteries in time, without the risk that the dose of medication is not administered to the patient.

In modern infusion pumps a sort of error log is available, which keeps track of how many times the unit has activated its alarm together with the activation causes. It is an excellent means for analyzing the different types of errors that can be committed and their frequency into the various wards. The classification of alarms is essential in order to be able to distinguish the cases where it is necessary to intervene immediately. As it can be seen from the graph in Fig. 9, the largest number of files contained in these error logs was found in oncology.

However, the efficiency of these alarm systems also depends on the operators intervention to timeliness, which depends primarily from experience. It was

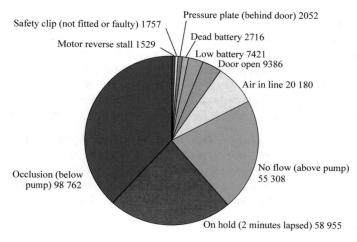

Fig. 8 Top 10 alarm codes

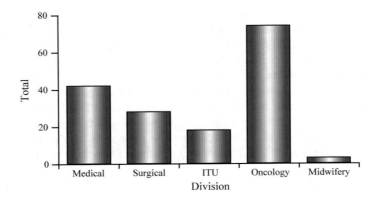

Fig. 9 Number of error log files by area

estimated that, considering an average intervention time of 5 min, corresponding to about 6% of the total time programmed for the infusion, the costs of maintenance and restoration of normal conditions are approximately £1000 per pump per year.

5   Critical Issues in Management and Safety

The issues about the safety of these devices generally find a large and complete response in the alarm system provided by the Manufacturer.

The reality shows a context where the use of medical equipment on the market still shows concerns about proper operation arising in the post-market phase and could impact on safety of patients and clinical staff. An excellent view on this phenomenon are the recall documents that the manufacturers publish in the Medical Devices Vigilance System, where the problems found during the market surveillance are reported, analysed and described, together with their fixing strategies.

A thorough analysis of these documents is certainly beyond the scope of this chapter. It is worth underlining that this panorama embodies many problems with a minimum impact on safety, easily solvable with information (user manual updating, integration of education and training). However, the number and the variability of these problems, particularly evident for infusion pumps, justify two critical issues:

- The need for a more accurate control both by technical personnel and by the user; these elements are certainly present in a correct model of management and maintenance of a medical device, requiring appropriate organizational efforts;
- Failure of the pre-market path and of the conformity control mode in ensuring, by themselves, the effective intrinsic safety of the device before release.

As an example, in the following is shown a safety notice from March 2013 (Q. FA.EMEA.2013.009) [8].

Issue: The problem highlighted is that, under particular circumstances (e.g. infusion is at a rate lower than a fixed threshold) the motor assembly may rotate backwards (roll-back) causing an over-infusion by capturing additional medication. As an additional critical element, it is noticed that the observed pumps may not detect this malfunction at flow rates of less than 2.0 ml/h, therefore not raising an alarm and causing over-delivering.

In this case the alarm system, to which the operator correctly relies on—even if sometimes with excess of confidence in the intelligence of the control systems—is another element of crisis in the security. An alarm could indeed be triggered during the power on self-test, or during the infusion process. In this case the start of a new infusion will be prevented or the possible ongoing infusion will be stopped, causing delays or interruption of the therapy.

Risk to Health: the severity of the case emerges immediately by reading this section of the safety notice. It is reported that, in the worst case and depending on the medication and dosage, the over-infusion could result in significant injury or death. Conversely, the delay or interruption of the therapy, or under-infusion, could as well cause significant injury or death.

What is reported in the remaining part of the report, in addition to highlighting the importance of the incident, is also especially relevant from the point of view of proper "culture" of security and co-operation with the professionals involved in the process of use. Health professionals are in fact asked to assess the risks/benefits associated with the use of the device for patients undergoing intensive therapy. It would be also appropriate considering the use of a different pump, particularly for patients for whom a delay or discontinuation of therapy could be a serious threat to health or death. The use restrictions are indeed heavy, since it is told not to use the pump throughout the neonatal and paediatric population up to two years and on any patient when you request a flow rate of 2 mL/h or lower.

6 Conclusions

As already mentioned, the modern infusion pumps, also called "smart pumps", are used in combination with software that allows an analysis of recorded errors. In addition to this function, programmed drugs libraries are available, in order to administer the right dose and concentration. Moreover, each patient has a barcode; corresponding to the drug he must receive, so as to facilitate the work of the nursing staff. However, despite these technological innovations, errors persist, although less frequent. In Massachusetts it was found that the so-called ADE (Adverse Drug Events) decreased by more than 70% over 16 months after the introduction of smart pumps [1]. In 2012 ten hospitals in the USA were the objects of an investigation aimed at identifying the most common errors associated with the use of smart pumps: labeling errors, unauthorized medications, incorrect identification of patients, wrong liquid dose and flow. 1691 errors were recorded on 1164 infusions, and in 60% of cases more than one error occurred in the same unit [9]. Most of these types of errors are related to a violation of hospital policy, which should regulate the management and use of different equipment through written instructions, in order to ensure their safety. However, often these written orders do not exist or they are not followed because the operations to be performed are not considered dangerous, but easily manageable arbitrarily. For example, in 87% of cases there is not a document that establishes the recommended intensity of the flow that serves to keep the patient line opened (KVO = Keep Vein Open), so it is the staff itself that sets the correct value considered, according to its knowledge. To make sure that these situations do not happen, it is necessary to intervene with adequate training of staff who will use these devices, always accompanied by activities of maintenance and upgrade (for example of drugs libraries). Also it is necessary to introduce infusion pumps into the legal metrology of each country and measure flow annually with established traceability [10–12].

The Clinical Engineering Division of the International Federation for Medical and Biological Engineering (IFMBE/CED) has started an E-Course project in 2016. The aim of this project is to train people living in developing countries on the activities involving medical equipment maintenance and management, using a virtual learning system for the lectures. The main expected outcomes for such a project are to develop a strategy that can support training of Clinical Engineers as well as Biomedical Equipment Technicians. A system that can, at low price, develop training courses in several parts of the world, using distant and local expertise and is not limited to language barriers [13, 14].

However, even the designers of smart infusion pumps must make sure to create products with functions that can be easily understood and used by the operators. Only thanks to the collaboration between manufacturers, technicians, doctors and nurses it becomes possible to organize the maintenance of any device, infusion pumps included. This not only means to check that they work properly according to the standards, but also to do analysis of the features and the functionality in the field, and to make changes if the performances are not those requested.

References

1. Breland BD (2010) Continuous quality improvement using intelligent infusion pump data analysis. Am J Health Syst Pharm 67:1446
2. ECRI Institute (2015) Top 10 health technology hazards for 2016. Report from "Health Devices", November 2015
3. EN 60601-2-24: 2015-05, "Medical electrical equipment Part 2: Particular requirements for basic safety and essential performance of infusion pumps and controllers"
4. IEC 60601-1: 2007-05, "Medical electrical equipment, Part 1: General requirements for basic safety and essential performance"
5. EN ISO 14971: 2012, ISO 14971, "Medical devices. Application of risk management to medical devices"
6. Food and Drugs administration website. http://www.fda.gov/MedicalDevices/ProductsandMedicalProcedures/GeneralHospitalDevicesandSupplies/InfusionPumps/

7. Lee PT, Thompson F, Thimblebly H (2012) Analysis of infusion pump error logs and their significance for health care. Br J Nurs 21(8) (Intravenous Supplement)
8. "Urgent field safety notice", Q.FA.EMEA.2013.009, https://www.swissmedic.ch/recalllists_dl/07515/Vk_20130403_15_e1.pdf. Accessed 22 Sept 2016
9. Schnock KO et al (2017) The frequency of intravenous medication administration errors related to smart infusion pumps: a multihospital observational study. BMJ Qual Saf 26 (2):131–140
10. Badnjevic A, Gurbeta L, Boskovic D, Dzemic Z (2015) Measurement in medicine—past, present, future. Folia Medica Facultatis Medicinae Universitatis Saraeviensis Journal 50 (1):43–46
11. Gurbeta L, Badnjević A, Inspection process of medical devices in healthcare institutions: software solution. Health Technol 7(1):109–117. doi:10.1007/s12553-016-0154-2
12. Gurbeta L, Alic B, Dzemic Z, Badnjevic A (2018) Testing of infusion pumps in healthcare institutions in Bosnia and Herzegovina. In: Eskola H, Väisänen O, Viik J, Hyttinen J (eds) EMBEC & NBC 2017. EMBEC 2017, NBC 2017. IFMBE Proceedings, vol 65. Springer, Singapore
13. IFMBE Clinical Engineering Division Website, http://cedglobal.org/. Accessed 29 Sept 2016
14. International Federation for Medical and Biological Engineering website, http://2016.ifmbe.org/organisation-structure/divisions/clinical-engineering-division/. Accessed September 2016

Cost Effectiveness and Increasing of Accuracy of Medical Devices In Legal Metrology System: Case Study Bosnia and Herzegovina

Dijana Vuković, Almir Badnjević and Enisa Omanović-Mikličanin

Abstract Appropriate treatment of patients largely depends on the correct and accurate diagnosis of medical experts. When diagnosing diseases and treating patients, medical personnel rely on results from tests received using various medical devices. Medical devices are classified as products of special importance for retaining the health of many people, and as such are subject to numerous regulatory investigations that determine the entire life-cycle of the devices, from production, sales, use and disposal. The aspect of safety, accuracy and precision, that is, the functionality of medical devices, is becoming increasingly important given that during the use of medical devices various factors lead to the degradation of performance. In this case, a strict, professional and independent inspection of the functionality of medical devices is of the utmost importance for ensuring precise diagnostics and treatment of the patient. The safety aspect of medical devices in health care is regulated worldwide by various agencies or by the application of international standards in healthcare facilities that ensure that the functionality of medical devices is checked at least once a year. Also, the introduction of medical devices into legal metrology in all countries of the world has made great strides in increasing safety, accuracy and quality of medical services, resulting in optimization of maintenance costs for medical devices. The aim of this study, with the help of statistical and economic methods, is to analyse the connections between a number of medical devices over which periodic inspection was initiated with increasing accuracy and safety of them, and to present and quantify the cause-effect reaction in the form of cost reduction for medical institutions.

D. Vuković (✉)
Faculty of Economics, University of Bihac, Bihać, Bosnia and Herzegovina
e-mail: dijana.vukovic@unbi.ba

A. Badnjević
Medical Devices Verification Laboratory, Verlab Ltd Sarajevo, Sarajevo,
Bosnia and Herzegovina

E. Omanović-Mikličanin
Faculty of Agriculture and Food Science, University of Sarajevo, Sarajevo,
Bosnia and Herzegovina

© Springer Nature Singapore Pte Ltd. 2018
A. Badnjević et al. (eds.), *Inspection of Medical Devices*, Series in Biomedical
Engineering, https://doi.org/10.1007/978-981-10-6650-4_13

1 Introduction

Health care systems are very complex and consist of a large number of factors that have a great impact on providing quality services to the end user, the patient. Each health care system consists of medical institutions that, along with human resources and infrastructure, must possess necessary medical devices. The diversity and innovation of medical devices, as a result of the development of the field of biomedical engineering, certainly contributes significantly to the improvement of the quality and efficiency of health services, and their proper functioning is crucial for the correct diagnosis and treatment of patients.

The aspect of safety, accuracy and precision, that is, the functionality of medical devices is becoming more and more important given that during the use of medical devices, various factors lead to degradation of performance. In that case, a strict, professional and independent inspection of the functionalities of medical devices is of great importance for ensuring accurate diagnosis, treatment and treatment of the patient. The aspect of the safety of medical devices in health care worldwide is regulated by various agencies or by the application of international standards in healthcare institutions that ensure that the functionality of medical devices is checked at least once a year. Also, the introduction of medical devices into legal metrology in all countries of the world has made great strides in increasing the safety, accuracy and quality of the provision of medical services, resulting in the optimization of maintenance costs for medical devices.

The health care system in Bosnia and Herzegovina is very specific and complex. It consists of three levels and over 20 subindices including various ministries, agencies and other institutions. Each of these levels and subspecies has different laws and regulations when the health system is at stake. All this significantly contributes to the different quality of services provided and access to health services [1].

At the state level, in Bosnia and Herzegovina, there are institutions that concentrate on the quality and efficiency of the healthcare system. One of them is the Agency for Medicinal Products and Medical Devices of Bosnia and Herzegovina, which was established by the Medicinal Products and Medical Resources Act as an authorized body responsible for medicine and medical resources that are produced and used in medicine in B&H. The scope of the Agency in the field of medical resources[1] is broad and includes [2]:

a. retaining a register of medical resources for the territory of B&H;
b. retaining a register of manufacturers of medical resources for the territory of B&H;

---

[1]Medical resource are instruments, devices, materials and other products that are applied on people and which their basic purpose, determined by the manufacturer, do not achieve pharmacological, immunological or metabolic activities, but are used alone or in combination, including the software for proper use.

c. retaining a register of legal entities carrying out wholesale of medical resources for the territory of B&H;
d. issuing a certificate of registration in the register of manufacturers of medical devices;
e. issuing the certificate of entry into legal entities registers of manufacturers of medical devices;
f. issuing a certificate of entry into the legal entities registers that perform a wholesale of medical devices;
g. issuing a certificate of entry in the register of medical devices;
h. collecting, analyzing and reacting to undesirable occurrences during the use of medical devices, or materiovigilance tasks of medical devices;
i. participation in work related to the assessment of the harmonization and labeling of medical devices in B&H with harmonized European standards and technical regulations adopted on the basis of the Law on Technical Requirements for Products and Conformity Assessment;
j. inspecting the supervision of production and wholesale of medical devices, as well as legal entities that manufacture or import and trade a wide range of medical devices within the framework of issued permits;
k. organization of the information system on medical devices, including the establishment of a database of medical devices that are registered in the medical device register, data on legal entities that produce medical devices, or import and wholesale many medical devices, collection of data on traffic and consumption of medical devices, data that enable the rationalization of the use of medical devices and the connection to international information networks of medical devices;
l. performing other activities in the field of medical devices in accordance with the Law, as well as regulations adopted on the basis of the Law.

The Medicinal Agency is a regulatory body in Bosnia and Herzegovina that regulates that any medical device placed on the market of Bosnia and Herzegovina is offered in accordance with the applicable norms and standards. This regulatory body has no defined rules for checking the accuracy of medical devices when they are installed to the end user.

As medical devices can be classified as measuring devices, one of the institutions directly involved in the regulation of the healthcare system, and within its capabilities contributes to its quality, is the Institute of Metrology (IMBIH), which is an integral part of the European Association of National Metrology Institutes (EURAMET). It has legal jurisdiction over the measurement institution. Among other things, IMBIH has the task of [3]:

- prepares draft laws and other regulations from the domain of its competence and coordinates the adoption of regulations for the work of metrology institutions within the entity,
- realizes the etalon base at state level and supervises the realization of the standard base of secondary and working etalons,

- ensures the traceability of measurements in B&H to international standards units,
- establishes, appoints and supervises metrology laboratories,
- supervises and coordinates the work of the control units for precious metal articles,
- perform calibration and inspection of standards, benchmarks and equipment used by institutions in metrology entities in accordance with the regulations in line with the recommendations and documents of the International Organization for Legal Metrology (OIML) as well as with other relevant international normative documents,
- deals with research and development activities, proposes and defines the priorities of realization of development projects and participates in international projects,
- develops studies, development strategies, projects, analyzes and other tasks in order to build a metrology system in B&H and metrological laboratory infrastructure,
- implement international cooperation agreements in the field of metrology, and participates in the projects of international organizations and through them represent Bosnia and Herzegovina.

In regards to the objectives of EURAMET in the field of medical metrology, IMBIH conducted a study entitled "Measurement in medicine" in 2014 through all public and private health facilities including three clinical centers, 26 hospitals, 63 health centers and more than 320 private Institutions in Bosnia and Herzegovina with the aim of defining metrological legislations in the field of medicine in Bosnia and Herzegovina, as already defined for other existing measures and measures in Bosnia and Herzegovina. The study was conducted by national team appointed by the IMBIH. All health care institutions were asked to define electrical medical devices that they use by name of manufacturer and model. Based on results of processing collected data, the team from the IMBIH suggested which medical devices should be defined as legal metrology and become subject to regular inspection in accordance with the ISO 17020 standard. For every proposed medical device, it was necessary to precisely define which outputs must be annually inspected and the minimum characteristics of etalons that will be used in inspection. The results of that research and study have shown that currently in Bosnia and Herzegovina there are a number of medical devices aged over 20 years. In most cases, these medical devices are not subjects of any service, or any kind of inspection. Medical devices of newer generation, as shown in this study, went under authorized preventive and corrective maintenance in different ways. In certain health institutions, preventive services are scheduled annually, whilst in certain institutions four times per a year with an impressive impact on the budget of the health care institution and for the overall health care system. This acts as a type of monopoly [4]. As a part of preventive services, an authorized service center performs also certification of apparatuses. Certification process reports are usually a work order document. This document only reports the results of the certification:

whether the device passed or failed. The work order document contains neither any information about device output values measurement nor the reference to the certification standard. Based on all collected data, as well as on the basis of international standards for medical equipment and metrology, the IMBIH team proposed 10 different medical devices to be introduced into legal metrology. The proposed medical devices are used in critical patient care, and are an integral part of every intensive care unit, operating room, and emergency care center. The proposed devices that are introduced into legal metrology system of Bosnia and Herzegovina, as of 2015 are: ECG, defibrillator, patient monitor, infusion pumps, perfusors, respirator, anesthesia machine, dialysis machine, neonatal and pediatric incubator and therapeutic ultrasound [5]. For these groups of devices, periodic inspections are performed by notified laboratories and measurement results, as well as other documentation, are stored in an online software for tracking the inspection of medical devices in the healthcare system of Bosnia and Herzegovina [6].

In regards to the aforementioned text, in this chapter, apart from the analytical methods for assessing the individual increase in the efficiency of medical devices, the overall social effect of changes is considered. It is therefore important to evaluate objectively and analytically all benefits and possible costs. This chapter presents the results of a study that includes all aspects of increasing safety and efficacy that are directly related to the introduction of medical devices into legal frameworks, and indirect benefits arising from that process, and are related to the direct reduction of the costs of medical institutions. The analysis is based on data collected in the territory of Bosnia and Herzegovina during the period of 01.01.2015 until 31.12.2016. This was collected through online software to monitor the status of inspection of legal benchmarks in healthcare.

2 The Role of Inspection of Medical Devices in Increasing the Accuracy and Efficiency of Medical Services

This chapter outlines the role of inspection of medical devices to increase accuracy and efficiency in Bosnia and Herzegovina over a period of two years. Table 1 and Fig. 1 shows the structure of correct and faulty medical devices in 2015 and 2016. In the first year, 18% of the total number of medical devices that are obliged to have a functionality review once a year under the Law on Metrology of Bosnia and Herzegovina are inspected. The share of exact devices in the total number of inspected in 2015 was 89.20%, while the share of faulty was 10.80%.

In the second year, inspection, functionality and accuracy were measured in 34% of devices on the market. The share of exact devices was 95.10%, and the inaccurate 4.90%.

Figure 1 depicts the increase in the number of inspected medical devices and their increase, and their accuracy. It is estimated that by 2018, all medical devices in the market of Bosnia and Herzegovina will be inspected, which will positively

Table 1 Increasing the accuracy of medical devices in 2016 compared to 2015

Year	2015 (%)	2016 (%)
Correct device	89.20	95.10
Incorrect device	10.80	4.90
Inspected medical devices	18	34

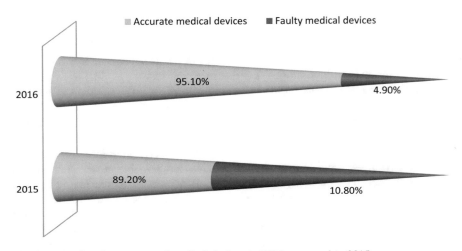

Fig. 1 Increasing the accuracy of medical devices in 2016 compared to 2015

reflect on the diagnostics, treatment and treatment of the patient. Ultimately, the goal of introducing medical devices into legal metrology is to remove all inaccurate devices from use so that the quality of the health service is at the highest possible level. In order to achieve this goal, it is necessary that all institutions involved in this process, from the perpetrator to the supervision, perform their work to full capacity.

2.1 Evaluating the Impact of Safety, Effectiveness and Costs in the Process of Legal Metrology Inspection of Medical Devices Using Data Envelopment Analysis (DEA)

Data Envelopment Analysis is a nonparametric method based on mathematics, more precisely linear programming, and is used for evaluating the relative effectiveness of comparable entities based on empirical data of their inputs and outputs [7].

Data envelopment analysis represents a set of models and methods which have foundation on mathematical programming. Data on selected *inputs* and *outputs* are included for all analyzed decision makers (DM) in linear program which represent the chosen model DEA. This way, it evaluates the efficiency of individual decision makers within a set of comparable decision-makers, that is, those who convert multiple *inputs* into multiple *outputs* identical to those of the observed decision-maker. Since the efficiency of each decision-maker is measured against other decision makers, its relative efficiency whose value is between 0 and 1, and deviations of 1, are attributed to the excess input or output deficit.

AOMP determines the empirical limit of efficiency (the limit of production possibilities) by limiting inputs from the bottom, and the output from above. Given that it is determined by the (best) existing decision makers, the efficiency barrier is a viable goal that should be pursued by inefficient decision-makers. They achieve efficiency by projecting to an efficient border. Thus, unlike typical statistical approaches based on average values, AOMP is based on extreme observations by comparing each decision maker only with the best ones.

The *input* and *output* weights are determined in such a way that each decision-maker joins a set of the most favorable weights. The term most favorable means that the resulting output and input ratio for each decision maker is maximized in relation to all other decision makers when those weights are associated with the respective *inputs* and *outputs* for each decision-maker.

The first step in modeling to be used is to assess the impact of the introduction of medical devices into legal metrology in Bosnia and Herzegovina and to identify the results (increasing the efficiency and safety of devices) that reflect the desired goals, as *inputs* which in this case are the percentage increase in the number of inspected devices. Choosing relevant inputs and outputs is one of the most important and, at the same time, the most difficult steps in the analysis that justifies the goal of conducting the analysis. *Inputs* and *outputs* should be selected so that *inputs* include all resources, and *outputs* all relevant activities or outputs for a particular efficiency analysis [8]. In addition, the ratio of the number of input and output variables and the number of units to be analyzed should be taken into account so that the results of the analysis are as close to reality as possible. Although there is no rule, the number of units should be at least 3 to 5 times greater than the total number of *input* and *output* variables.

Three indicators (two inputs and one output) were selected for the purpose of this paper. For each medical device that is under legal metrology, the underlying inputs are included in the analysis:

X1—the number of inspected medical devices in 2015 expressed as a percentage of the total number of devices on the market

X2—the number of inspected medical devices in 2016 expressed as a percentage of the total number of devices on the market

and one output: **Y1—increasing the number of correct medical devices in percentages,** where applicable (j = 1, 2,..............., 9).

Table 2 depicts the number of inspected medical devices in 2015 and 2016 expressed in percentages. It can be observed that there was an increase in the number of inspected devices from 2 to 59% in 2016 compared to 2015. If we observe the share of the total number of medical devices in the market of Bosnia and Herzegovina, then we can see that the most inspected are dialysis machines

Table 2 Increasing the accuracy of each inspected medical device in 2016 compared to 2015

	The percentage of inspected medical devices in 2015	The percentage of inspected medical devices in 2016	Percent point	2015		2016		Percent point
				X (%)	✓ (%)	X (%)	✓ (%)	
	X1 (%)	X2 (%)						Y1
Anesthesia machine	19	38	19	19	81	9	91	10
Defibrillator	27	42	15	6	94	3	97	3
Dialysis machine	10	69	59	18	82	1	99	17
ECG	29	39	10	7	93	4	96	3
Infusion pumps and perfusors	2	14	12	27	73	2	98	25
Neonatal and pediatric incubator	15	29	14	22	78	17	83	5
Patient monitor	12	47	35	12	88	7	93	5
Respirator	14	29	15	6	94	10	90	−4
Therapeutic ultrasound	27	29	2	30	70	7	93	23

with 69%, while the least-inspected are infusions pump and perfusors (14% of the market).

The largest increase in 2016 compared to 2015 had dialysis machines of as much as 59% points. When we observe an increase in the accuracy of medical devices, then we can say that the number of accurate medical devices in most cases increased in 2016 compared to 2015. If we increase of the accuracy of the device, put into the context of increasing the number of inspected devices on the market, then we can conclude that increasing the number of inspected medical devices has increased their accuracy. In the case of respirators, we have a situation that in the sample of 14% of these devices on the market, 6% were incorrect, while by increasing the sample by 15% points, the share of inaccurate devices increased to 10%.

The core models within the data sharing analysis, which are also commonly used, are the Charnes-Cooper-Rhodes (CCR) model and the Banker-Charnes-Cooper (BCC) model. The choice of the model does not depend solely on theoretical settings, but also on the context and purpose of the analysis, as well as on the long-term or the short-term consideration [9]. Generally, for basic data sharing analysis models (CCR and BCC models), there are certain general rules and assumptions. It does not require the input and output value to the same measurement units of the method to function equally using different measurement units, and this is one of its greatest advantages.

The CCR model assumes constant yields with respect to the scope of the action, and due to the modification of this assumption, other data limiting models have emerged including the BBC (Banker-Charnes-Cooper) model. The BBC model assumes variable yields on the scope of action, and the production boundary that is trampled with a convex shell of decision makers with linear and concave characteristics. In the case of rising or decreasing yields, in which the proportional increase in input results in more, or less than a proportional increase in output, the BCC model should be selected.

When choosing a type of model, the characteristics of the data and knowledge of the yield type characteristic of the analyzed process are decisive. Since the process of inspection of the medical devices is a process with variable yields (increasing the input does not necessarily mean a proportional increase in output), the BCC model is recommended. Namely, if the input is increased, i.e. the number of inspected medical devices does not necessarily mean a proportional increase in the safety and efficiency of the device (higher or lower values are possible). The goal of medical devices inspection is to maximize the safety and efficacy, with the assumption of as many device inspections as possible, so a BCC model focused on output will be used.

The model has two inputs and one output, and the analysis is based on 10 medical devices that have entered the legal metrology of Bosnia and Herzegovina, with infusion pumps and perfusors being observed together. Consequently, it is possible to conduct a data limiting analysis (the rule is $3 \times (2 + 1) = 9$) because at least 9 devices are needed, which satisfies the starting models' assumptions. Data limiting analysis was performed using the software-a MaxDEA basic 7.7 (Table 3).

The table represents the aggregation data of the data limiting analysis, that is, the number of entities that entered the analysis, where the output-oriented model has two inputs and one output (Table 4).

The results obtained by the data-limiting analysis, more precisely by the BBC-based exit-oriented method, show that the inspection of medical devices: infusion pumps and perfusors, achieved the highest efficiency, i.e., the increase in the inspected sample increased the accuracy of the device, i.e. the efficiency index was 1 (100%), which means that the relative impact of the increase inspected devices to inflate accuracy are extremely high.

Table 3 DEA model—Results summary

Property	Value
Model type	Envelopment model
Number of DMUs	9
Number of inputs	2
Number of outputs	1
Distance	Radial
Orientation	Output-oriented
Returns to scale	Variable
Slack computation	1 Stage
Elapsed time	3 s

Table 4 Results of
envelopment model

DMU	Score
Infusion pumps and perfusors	1
Therapeutic ultrasound	0.92
Dialysis machine	0.68
Anesthesia machine	0.4
Neonatal and pediatric incubator	0.2
Patient monitor	0.2
ECG	0.12
Defibrillator	0.12
Respirator	0

The smallest relative effect on increasing the accuracy by increasing the number of inspected devices is according to the DEA method with the respirator.

2.2 Analysis of Accuracy and Performance Inspection Process During 2015 and 2016

This part of the study analyzes the increase in the efficiency and accuracy of all ten medical devices that entered the legal metrology in Bosnia and Herzegovina individually. As mentioned before, proposed devices are: anesthesia machine, defibrillator, dialysis machine, ECG, infusion pumps, perfusors, neonatal and pediatric incubator, patient monitor, respirator and therapeutic ultrasound. Previously, it was shown that at the aggregate level, the accuracy of medical devices increased with the increase in the number of inspected devices, and the changes that are affected are detailed below. Such an analysis is extremely important for the users of medical services in Bosnia and Herzegovina as every device removed from use reduces the potential consequences for health and ultimately leads to a more efficient and safer health system.

Figure 2 represents the correctness of the inspected Anesthesia machines in 2015 and 2016. The illustration shows that in 2015, as much as 19% of inspected devices were faulty, while in 2016, by increasing the share of inspected Anesthesia machines, the faulty devices in the market of Bosnia and Herzegovina were reduced. Therefore, we can conclude that the quality and efficiency of medical services involving this type of device has increased (Fig. 3).

Although a single faulty medical device can cause enormous losses and consequences for both the patient and the medical system as a whole, it should be emphasized that, unlike Anesthesia machines, Defibrillators in both observed years had a relatively small fraction of faulty devices in the structure of the total number of these devices on the market. Namely, in 2015, there were 6% of faulty devices with a decrease in tendency (3%) in the market in 2016 (Fig. 4).

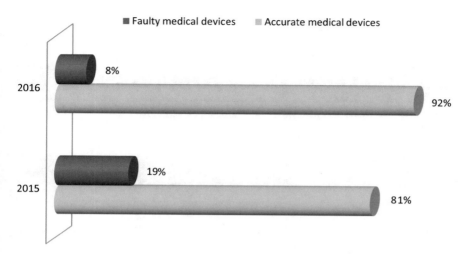

Fig. 2 Increasing accuracy and performance of Anesthesia machine in 2016 compared to 2015

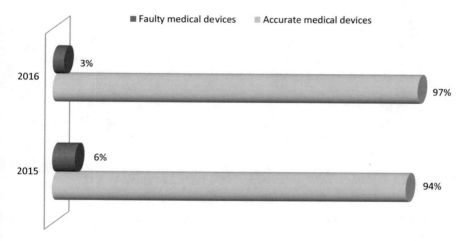

Fig. 3 Increasing accuracy and performance of Defibrillator in 2016 compared to 2015

Outstanding results were achieved by reviewing Dialysis machines in 2016 compared to 2015. In 2015, the share of faulty devices were as high as 18%. In just one year and over 69% of the market, inaccuracy of Dialysis machines was reduced to 1%. From the above it can be concluded that the effect of inspection in the first year of application of the new regulation was extremely high when this type of device was at stake (Fig. 5).

Regarding the ECG device in terms of the percentage of correct devices in 2015, the market of Bosnia and Herzegovina was 93%, while the share of faulty was 7%. Compared to 2015, the correctness of the ECG device had increased to 96%, and

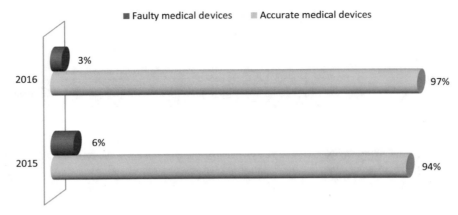

Fig. 4 Increasing accuracy and performance of Dialysis machine in 2016 compared to 2015

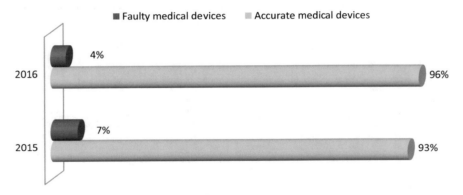

Fig. 5 Increasing accuracy and performance of ECG in 2016 compared to 2015

4% of the device had not passed inspection and had to be permanently removed or repaired (Fig. 6).

Early analysis has shown that the efficiency of Infusion pumps and perfusors inspection is high, which can be seen from the accompanying illustration. Namely, the share of faulty devices in 2015 from 27% decreased to 3%, observing the structure of the correct and faulty devices in the territory of Bosnia and Herzegovina in 2015 and 2016.

Figure 7 shows the structure of correct and faulty Neonatal and pediatric incubators. The inspection has reduced the share of faulty devices in the total number on the market, however, there is still a high rate of devices that did not meet standards, and additional efforts should be made to increase the safety and efficiency of these devices.

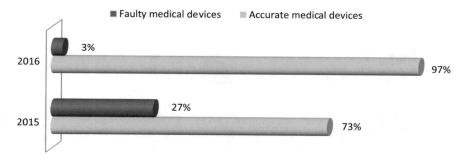

Fig. 6 Increasing accuracy and performance of Infusion pumps and perfusors in 2016 compared to 2015

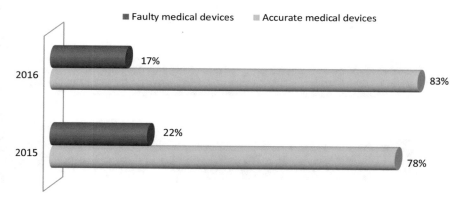

Fig. 7 Increasing accuracy and performance of Neonatal and pediatric incubator in 2016 compared to 2015

Figure 8 shows Patient monitors, i.e. the movement of the share of faulty devices in 2016 compared to 2015. From the depicted images, there is a visible increase in the correctness of the observed medical device by 5% (Fig. 9).

A unique case in the individual analysis of the devices introduced into the legal metrology in Bosnia and Herzegovina is in the direction of the movement of the share of faulty Respirators. Namely, in the sample which was examined in 2015, there was a smaller share of faulty devices than in 2016. The reason for this is simple, which is a statistical constraint and a possible unrepresentative sample (Fig. 10).

In 2016, with a large 30% of faulty Therapeutic ultrasounds, the proportion was reduced to 7% of the overall structure of the device in use. In 2016, with a large 30% of irregular Therapeutic ultrasounds, the proportion was reduced to 7% of the overall structure of the device in use.

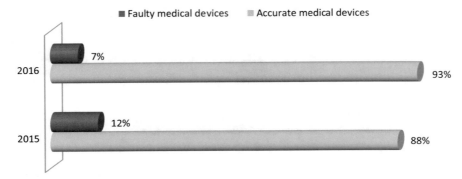

Fig. 8 Increasing accuracy and performance of Patient monitor in 2016 compared to 2015

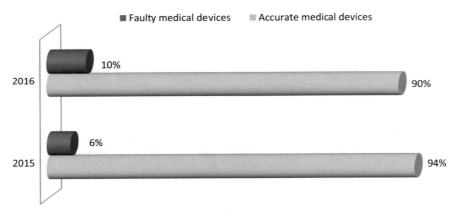

Fig. 9 Accuracy and performance of Respirator in 2016 compared to 2015

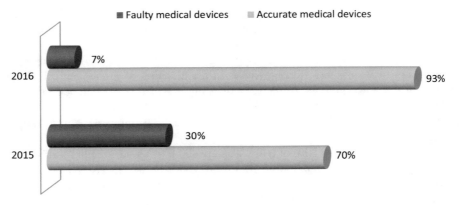

Fig. 10 Increasing accuracy and performance of Therapeutic ultrasound in 2016 compared to 2015

From the analysis carried out above, we can conclude that as a result of the introduction of medical devices into legal metrology and their inspection, there has been a significant increase in the efficiency and correctness of medical devices in Bosnia and Herzegovina, and that their users or patients are safer and that the quality of the health system is certainly higher.

3 Analysis of the Expenses and Savings in Order to Estimate the Cost-Effectiveness of the Introduction of the Medical Equipment into the Legal Metrology in Bosnia and Herzegovina

The assessment of the influence that the introduction of medical devices has in legal metrology is a procedure that requires evaluation of the acceptance of this kind of "project" with respect to accuracy, safety, and determination of the required measures of protection for the patient, which are determined by the law of Bosnia and Herzegovina.

Analysis of the benefits and expenses (with the analysis of other influences) should provide unanimous decision if the society is ready to accept (pay) the amount of expenses that certain procedures will cause for the society (or for the individual), compared to the benefits that will be provided by the procedure. Total benefits and drawbacks are based on possible direct and indirect influences, such as:

- human health
- ecosystem
- economy
- society

they should also consider all other relative factors, (measurable and nonmeasurable) as well as their influence and the possibility of their emergence. Final results represent the assessment of the societal benefits and expenses, or in the other words, gains and losses in social well-being when it comes to the realization of this procedure. This is why the main goal for the introduction of the medical equipment into the legal metrology is the increase of the usefulness for the users of the medical equipment, firstly for the safety and the protection of the patients, as well as for the reduction of the expenses for the maintenance of the medical devices, and the safety of doctors and institutions that own these devices. Annually inspected medical device allows the achievement of these set goals.

This chapter explores the validity of the introduction of the medical devices into the legal metrology in B&H with aspect to reduction of the expenses. Systems for the maintenance of the medical devices is a very complicated process that has been analyzed according to the international standards and guides, and it is a part of the risk analysis for every medical institution. The problem in the system of health protection in B&H has been solved by the introduction of periodical inspections, which provide additional safety and better service quality, and are preventive and allow planning required for the replacement/procurement of new parts/new devices. This prevents placing key medical devices out of use, which could lead to a halt in providing medical services to the patients and cause patient's dissatisfaction, delay in medical service, uneven schedule of work load for medical experts, and so on. In order to achieve set goals, it is important to review the expenses for the maintenance of the medical devices before and after of the introduction of the amended Law about metrology [10].

Also, it is important to estimate the effects on the wider social community and medical institutions. It is important to emphasize that up until the change of this law, preventive and corrective check-ups of the medical devices were conducted based on the discretion of the agency for the medical devices, and that corrective interventions were much more common than preventive ones. That kind of inspection brings the performance of the device into question as there is no guarantee that the device was working in optimal conditions, and this in the end increased the risk that the patient received inadequate therapy (lower quality therapy).

Table 5 shows expenses before and after introduction of medical devices in legal metrology in Bosnia and Herzegovina. As shown, total expenses for the maintenance of the medical devices, which by law are required to be inspected, are reduced by 83% in the case of the inspection of all devices.

When we analyze the data from 2015 and 2016, that is, in the analysis, we include only the number of inspected devices by the specified years and we multiply with the maintenance of prices and prices after being introduced into legal metrology, then we can say that the costs have decreased by 87% in 2016 compared to 2015. And, without any doubt, confirm that the introduction of new regulations is gaining more benefits in order to increase safety, correctness and accuracy, as well as reducing the costs of medical institutions. Namely, funds that are freed from cost reduction can be directed at improving the quality of services and creating a multiplication of benefits for the health system as a whole.

If we observe a total reduction in costs for each medical device introduced into legal metrology in Bosnia and Herzegovina (Fig. 11), we can notice that much smaller financial resources are allocated for the maintenance of all ten devices, that is, achieving extremely high savings. Savings range from 50% (Infusion pumps and Perfusors) to 88% for the Anesthesia machine. A high percentage of anesthesia machine are achieved due to the fact that they are mostly maintained by authorized service providers who have changed parts of the device by the manufacturer's recommendation, e.g. recommendations for replacing O2 sensors are once a year and in most cases, replacements are made every 6 months, regardless of the frequency of the use of the device. Also, preventive examinations are conducted by a service technician without the impact of a medical device maintenance service or an external device performance appraiser.

Table 6 shows the total savings per year for individual medical institutions owned by the observed medical devices.

Table 5 Overview of costs savings before and after the implementation of the law on metrology of medical devices

Medical devices	The percentage of inspected medical devices in 2015 (%)	The percentage of inspected medical devices in 2016 (%)	Costs savings in 2016 compared to 2015 (%)	The percentage of inspected medical devices in 2018 (Projection) (%)	Costs savings in 2018 compared to time before implementation of the Law (%)
Anesthesia machine	19	38	100	100	88
Defibrillator	27	42	53	100	82
Dialysis machine	10	69	568	100	68
ECG	29	39	36	100	84
Infusion pumps and Perfusors	2	14	482	100	50
Neonatal and pediatric incubator	15	29	97	100	83
Patient monitor	12	47	300	100	77
Respirator	14	29	106	100	77
Therapeutic ultrasound	27	29	5	100	60
Total	18	34	87	100	78

As the analysis showed, a great reduction in costs has been achieved for all medical institutions. The largest savings were recorded in health centers in the amount of 81.6%, followed by hospitals (79.3%), private medical institutions (76.8%), and at least in clinical centers (74.7%). The reasons for this are the complexity of the listed institutions and the number of appliances they have. In the database that analyzed the least data, there are devices from clinical centers, so these results are a relative indicator, but it is interesting that the results of the analysis on a small number of devices from clinical centers show a similar trend which means that the same problem is present throughout the health system from the lowest instances to the highest level.

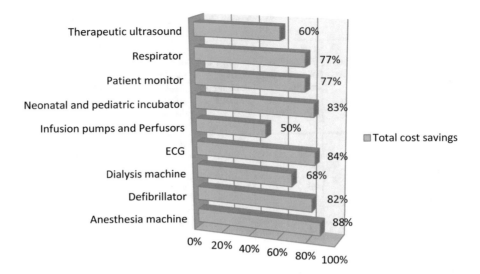

Fig. 11 Total costs savings per medical devices

Table 6 Overview of cost savings by medical institutions

No.	Type of institution	Total% savings after applying the law on an annual basis (%)
1	Clinical centers	74.7
2	Hospitals	79.3
3	Health centers	81.6
4	Medical private institutions	76.8
	Total	78.0

4   Conclusion

The primary role of each health care system is to provide an effective, parametric accurate, safe and equal service to all patients. The introduction of medical devices into legal metrology is one of the most important steps for regulation and standardization of the health system in Bosnia and Herzegovina. The analysis has shown that periodic inspections of medical devices achieve phenomenal results in increasing the correctness and efficiency of medical devices, which is extremely important for patients and the medical system as it contributes to increasing the quality of service provided. Consequently, periodic inspections have enabled the collection of a large database that consist of basic information about devices such as manufacturer, type and age of the device, as well as the performance of medical devices. This database represents an exceptional starting point for risk analysis, which is an integral part of the management of each healthcare institution in the world, planning the replacement of medical devices with respect to performance degradation, and planning and optimizing corrective examinations of medical devices. Also, given the existence of a digital database of measurements and accompanying documentation, healthcare institutions are no longer burdened with a large number of documents for each device, and a move towards the introduction of information systems for the management of medical devices covered by this regulation, which is an already established practice in healthcare systems of developed countries in the world.

In the end, in addition to the benefits mentioned above, the analysis also revealed an extremely significant reduction in costs for medical institutions. Namely, at the time of general illiquidity of the market, such as in Bosnia and Herzegovina, for institutions possessing the mentioned devices, it is extremely important to achieve as much savings as possible but with the increase in the safety and quality of their services. Thus, by introducing medical devices into legal metrology, win-win positions are achieved where, with the increase in the accuracy, safety and quality of services, great savings are formed in the healthcare system. In addition to the savings that have been achieved, the satisfaction of medical experts in the health-care system in Bosnia and Herzegovina is also welcomed, welcoming the introduction of this legislation due to the certificate of conformity/mismatch between medical devices and the avoidance of situations in which staff claims that the device is not correct, and the maintenance service claims the opposite.

References

1. Mujkić E (2015) Sistem zdravstva u Bosni i Hercegovini: stanje I pravci moguće reforme. Fondacija Centar za javno pravo [last accessed on 23rd April, 2015]
2. Zakon o lijekovima i medicinskim sredstvima, Službeni glasnik BiH, br. 58/08
3. Zakon o mjeriteljstvu, Službeni glasnik BiH, broj 19/01
4. Badnjević A et al. (2015) Medical devices in legal metrology. Embedded Computing (MECO), 2015 4th Mediterranean Conference on IEEE
5. Badnjević A et al. (2015) Measurement in medicine—past, present, future, Folia Medica Facultatis Medicinae Universitatis Saraeviensis 50.1
6. Gurbeta L, Badnjevic A (2016) Inspection process of medical devices in Healthcare Institutions: software solution, Springer Health and Technology Journal
7. Rabar D (2010) Ocjenjivanje efikasnosti poslovanja hrvatskih bolnica metodom analize omeđivanja podataka, Ekonomski pregled 61, 9–10, Zagreb
8. Rowena J, Smith PC, Street A (2006) Measuring efficiency in health care: analytic techniques and health policy. Cambridge University Press, Cambridge
9. Rowena J, Smith PC, Street A (2006) Measuring efficiency in health care: analytic techniques and health policy. Cambridge University Press, Cambridge
10. Zakon o mjeriteljstvu, Službeni glasnik BiH, broj 19/01

Printed in the United States
By Bookmasters